Boost Your STEAM Program
with Great Literature
and Activities

BOOST YOUR STEAM PROGRAM WITH GREAT LITERATURE AND ACTIVITIES

Liz Knowles, EdD, and Martha Smith

LIBRARIES
UNLIMITED™

An Imprint of ABC-CLIO, LLC
Santa Barbara, California • Denver, Colorado

Library of Congress Cataloging-in-Publication Data

Names: Knowles, Elizabeth, 1946- author. | Smith, Martha, 1946- author.
Title: Boost your STEAM program with great literature and activities / Liz
    Knowles, EdD, and Martha Smith.
Description: Santa Barbara, California : Libraries Unlimited, an imprint of
    ABC-CLIO, LLC, [2018] | Includes bibliographical references and index.
Identifiers: LCCN 2018003077 (print) | LCCN 2018014862 (ebook) | ISBN
    9781440862519 (ebook) | ISBN 9781440862502 (paperback : acid-free paper)
Subjects: LCSH: Science–Study and teaching–Bibliography. |
    Science–Juvenile literature–Bibliography. | Art–Study and
    teaching–Bibliography. | Art–Juvenile literature–Bibliography. |
    Picture books for children–Bibliography. | Interdisciplinary approach in
    education–Bibliography. | Project method in teaching–Bibliography. |
    Activity programs in education–Bibliography.
Classification: LCC Z5818.S3 (ebook) | LCC Z5818.S3 K66 2018 LB1585 (print) |
    DDC 028.5/34–dc23
LC record available at https://lccn.loc.gov/2018003077

ISBN: 978–1–4408–6250–2 (paperback)
        978–1–4408–6251–9 (eBook)

22 21 20 19 18   1 2 3 4 5

This book is also available as an eBook.

Libraries Unlimited
An Imprint of ABC-CLIO, LLC

ABC-CLIO, LLC
130 Cremona Drive, P.O. Box 1911
Santa Barbara, California 93116-1911
www.abc-clio.com

This book is printed on acid-free paper ∞

Manufactured in the United States of America

CONTENTS

INTRODUCTION

As educators, our attention has recently been focused on an approach to teaching and developing curriculum that endorses the old-time concept of interdisciplinary curriculum development and teaching. The most important element is that STEAM includes K-5 students! In order to really embrace the process, collaboration across subject areas is imperative. What better place to start it but in the elementary grades? Collaboration across the middle school grade levels and subject areas was imperative to produce an effective interdisciplinary unit many years ago. Collaboration is required to do an effective job of curriculum mapping and then use the maps to develop cross-curricular themes across grade levels and subject areas. And collaboration is at the core of any successful STEAM program.

The other important thing that has come out of this is the freedom for students to problem-solve without fear of failure. Part of this is to let students know that failure is almost expected—that they have to try, try, and try again to solve a problem. Students learn that sometimes failure leads in a different direction and to a successful but different result/solution. It is and will be difficult for teachers to let go of precise and detailed step-by-step directions, benchmarks and rubrics, and the acceptance of the one and only correct answer. And the whole concept of makerspace is fascinating—like the 5th-grade science fair project on steroids and for all ages. It also touches on the entrepreneurial spirit of kids—it is not difficult to find kids who have gone way beyond the lemonade stand!

Educating students to traditional literacy standards is no longer enough. Independent and creative thinking are highly valued if students are to thrive in their academics and eventually their 21st century careers. Students must master a completely different set of skills to succeed in a culture of technology-driven automation, abundance, and access to global labor markets.

This new emphasis on digital literacy also calls attention to skills that have not been focused upon before. These skills include online research—knowing where to look for information and then knowing how to discern if the information is accurate. It also requires access to experts in the field. It necessitates working together with classmates, communicating ideas, listening to, and contemplating and valuing the ideas of others.

In order to keep up with all the information available in educational journals, a very helpful resource is Kim Marshall's *Marshall Memo*—a subscription-based weekly newsletter delivered by email. Kim reads from a huge list of professional journals and publications and summarizes the articles that he thinks are most interesting and pertinent. He provides the link to the full article if you want to read more. This resource is immensely helpful and time saving. http://www.marshallmemo.com/

Another site that is an excellent resource for locating books about STEAM topics and information on STEAM authors is http://www.teachingbooks.net. Nick Glass does a great job of providing valuable materials for teachers and students about books, authors, and illustrators; the fee for access to the site is very reasonable.

In the *Educational Leadership* (2015) that is all about STEM, Sunday Cummins tells of the importance of "Reading about Real Scientists." The article describes informational materials based on the five specific subject areas that are included in STEAM. It is important for students to understand how inventors in the past developed the end results that they achieved; most often it was through trial and error. It is very helpful and affirming to read about scientists who lived long ago and how frustrating it was for them to not quite reach their goal. Sometimes, they invented something that they discovered on a tangent rather than actually solving the original problem. Being able to develop a general problem to solve and then find a set of reference books and materials that will support the knowledge base needed in order to create reasonable solutions is an important part of this entire movement. According to Cummins, an anchor text should serve as a starting point for developing ideas and gathering background information. There are many possibilities for using books and journals, magazines, videos, and original documents in order to support and add to possible solutions. We have created a resource that identifies a huge number of these "anchor texts" (Cummins, Sunday. "Reading about Real Scientists." *Educational Leadership* 72:4 [2014–2015]: 68–72).

In reviewing the research for this book, I provided the highlights of the articles and a few books so that readers could get the essence of the information regarding the STEM/STEAM movement and then decide if they wanted to locate the information to read in more detail. This was done because I know teachers do not have time to do research and read multiple journals. No thoughts in those review/summary paragraphs are mine. I have quoted or paraphrased the essence so that it would read exactly or very close. So—for each article or book—the bibliographical information is included to facilitate the location of the complete work, if desired.

Martha Smith and I put together this book that features some information from our previous titles and, most important, extensive nonfiction bibliographies for each of the five STEAM topics. We feel that this will serve as a much-needed resource for teachers, parents, and librarians who are introducing and building a STEAM program for students.

Chapter 1

WHY STEAM?

THE CASE FOR STEAM

- Leonardo Da Vinci was the first STEAM maker!
- STEAM programs highlight the 21st century skills of collaboration, communication, creativity, and critical thinking.
- Students are never too young for STEAM.
- Graduates of STEAM/STEM programs are highly sought after.
- STEAM is the integration of standards from the disciplines of science, technology, engineering, art, and math.
- Project-based learning (PBL) is the best instructional delivery system for STEAM.

There is currently a great demand for graduates into the workforce who have STEAM/STEM skills. Computer programming and IT jobs are the hardest jobs to fill according to some recent studies. Despite this information, computer programming and STEAM-related programs in colleges are underserved.

The addition of arts to the STEM program has made it more accessible to students who were not comfortable with a focus on math and science. Students might be more engaged if it includes something they are familiar with. The focus on integrating the five subjects in a STEAM program helps the students to realize that the subjects are not segregated in the real world. They all blend together and skills from all five areas are needed in the workplace and really in everyday life.

Now if a student has been uncomfortable with math and if math is now a part of a project that includes art and technology, then it is not as overwhelming. With art as part of the program, girls might be more likely to want to participate. The major components of STEAM programs are critical thinking and problem-solving, so it is important to introduce those skills at an early age. Art can be used to sketch ideas and plans, coding can be embedded in the creation of videos, and languages can be practiced when the student is asked to make the final presentation about the project in Spanish.

The future job market is not to be denied. The constantly changing world makes it difficult for us right now to know what is ahead for the students, but we do know that the skills of collaboration, critical thinking, communication, and creativity are going to be needed in the workplace in the future regardless of the type of job.

STEAM provides the freedom for students to problem-solve without fear of failure. Part of this is to let students know that failure is almost expected—that they must try, try, and try again to solve a problem. Students learn that sometimes failure leads in a different direction and to a successful but different result/solution. It is and will be difficult for teachers to let go of precise and detailed step-by-step directions and the concept of a one and only correct answer. And the whole concept of makerspace is fascinating—like the 5th grade science fair project on steroids and for all ages. It also touches on the entrepreneurial spirit of kids—it is not difficult to find kids who have gone way beyond the lemonade stand!

ANNOTATED JOURNAL ARTICLES

Cummins, Sunday. "Reading about Real Scientists." *Educational Leadership* 72:4 (2014–2015): 68–72.

> Using series books such as The Scientific Mystery Series from Millbrook Press and the Scientists in the Field Series from Houghton Mifflin Harcourt Books for Young Readers, the author developed a program for students to use titles from these series to learn more about what scientists and inventors do. It is important for students to understand how inventors in the past developed the end results that they achieved; most often it was through trial and error. Sometimes, they invented something that they discovered on a tangent rather than solving the original problem. Being able to develop a general problem to solve and then find a set of reference books and materials that will support the knowledge base needed to create reasonable solutions is an important part of this entire program. There are many possibilities for using books, journals, magazines, videos, and original documents to help students truly understand what scientists and inventors do and for information needed to solve a current dilemma.

Wright, Tanya and Amelia Wenk Gotwals. "Supporting Disciplinary Talk from the Start of School: Teaching Students to Think and Talk Like Scientists." *The Reading Teacher* 71:2 (2017): 189–197.

> Teachers need to encourage reading, writing, and talking about scientific topics from the early grades. As early as kindergarten, they should use driving questions and develop scientific vocabulary. It is important that they investigate how things work and grow. Teacher read-alouds are an integral part of this—quality nonfiction books should be the focal point of scientific investigations. It is also important to have a conversation about learning to put things in perspective.

Gorman, Michael. "What Is STEM Noun or Verb? STEM in all Areas.... Ten Ideas to Transform STEM from Nouns to Verbs ... and Facts to Thinking." 21st Century Tech.

September 26, 2016. https://21centuryedtech.wordpress.com/2016/09/26/stem-for-in-all
-areas-ten-ideas-to-transform-stem-from-nouns-to-verbs-and-facts-to-thinking/
> The author shares some important ideas in this article. Highlights include the focus on skills
> needed for careers of the future—regardless of what that future holds—the basic skills
> of critical thinking, communication, creativity, and collaboration will prevail. It is important
> for students to see the real-world connection to what they are doing and learning. Rubrics
> need to focus on the process, not the product. It is important that students are "doing" and
> that they are also thinking about what is happening and applying necessary skills.

Stager, Gary S. "What's the Maker Movement and Why Should I Care?" *Scholastic Administrator*. Winter 2014. http://www.scholastic.com/browse/article.jsp?id=3758336
> The author states that students can be amazing thinkers, designers, and problem-solvers.
> Within a makerspace, they can utilize new technologies to create in surroundings where
> they can try and try until they are satisfied with the results. Teaching and learning are not
> synonymous—we should never expect that because we teach, they learn! Understanding
> core scientific ideas is best represented by the application of those ideas to solve problems.
> Once using the makerspace to create and problem-solve, they will be able to transfer that
> to their surroundings and the real world and hopefully make a difference.

Soule, Helen. "Why STEAM Is Great Policy for the Future of Education." *P21.org*. March 31, 2016. http://www.p21.org/news-events/p21blog/1900-why-steam-is-great
-policy-for-the-future-of-education
> A similar thought is shared by Soule who states that once students have been able to cre-
> ate and problem-solve in a makerspace, they develop their creativity and critical thinking
> skills so that they can apply concepts already learned in content classes in useful and pro-
> ductive ways (Soule 2016). Being able to think like a scientist or an engineer can have
> powerful results in any career.

Berg, Brenda. "The Critical Importance of STEAM Education." *PC World*. June 27, 2017. https://www.pcworld.idg.com.au/article/621170/critical-importance-steam-education/
> One of the main goals of education should be to make sure that students are prepared
> for careers that they will be going into upon graduation. With advances in technology,
> engineering, and design, we know that a STEAM education helps students to think
> critically and creatively and become independent thinkers and apply those skills to
> problem-solving.

Newell, Jennifer. "Girls, Computers, and STEAM." *BEANZ*. July 31, 2017. https://
www.kidscodecs.com/girls-computers-steam/
> There needs to be continued focus on presenting the field of computer science to young
> girls so they can see the importance of technological skills in creativity and problem-
> solving for the future.

Stewart, Louise. "Maker Movement Reinvents Education." *Newsweek*. September 8, 2014. http://www.newsweek.com/2014/09/19/maker-movement-reinvents-education
-268739.html
> This article features the ideas of Tony Wagner, current expert-in-residence at Harvard
> University's new innovation lab. He says that school should no longer be a place where
> knowledge is transferred from teacher to student and that is the end of it. Being able to
> remember and repeat things learned for success on a test is not the appropriate

preparation for the real world. He says that trying and failing and trying again is an idea that is not reinforced in our education system. However, working in a makerspace allows a student to take an idea and try to implement it over and over again until the results are better and something amazing has been achieved. It is important that we prepare students to look at all problems as solvable.

Bajarin, Tim. "Maker Faire: Why the Maker Movement Is Important to America's Future." Time. May 19, 2014. http://time.com/104210/maker-faire-maker-movement/

The maker movement is important because it has the potential to turn students into creative thinkers who will apply that mindset to the real world in their careers. If we provide them with the space, the tools, and the support they need to let their imaginations run wild, we can be confident that they will change the world, one problem solved at a time.

Berkowicz, Jill and Ann Myers. "How a STEM Shift Makes Way for Equity." Corwin Connect. November 29, 2016. http://corwin-connect.com/2016/03/stem-shift-makes-way-equity/

These authors relate that the opportunities provided by a STEAM/STEM education will help all students—in all grade levels and all types of learners. Creating and thinking to solve problems will entice all students to take part and grow in this learning environment. Anyone can be part of the excitement, part of the team, and participate in curiosity, collaboration, and discovery. Students at all levels will be energized by applying ideas and knowledge to a problem and then persevering to arrive at a workable result. This is a huge opportunity for schools to get on board with those 21st century skills of collaboration, creativity, critical thinking, and communication—all valued and necessary in the future world of the students.

Vasquez, Jo Anne. "STEM—Beyond the Acronym." *Educational Leadership* 72:4 (2014–2015): 10–15.

STEAM was developed to remove the barriers between the content areas and the way they are taught—as separate, unrelated entities. It is sometimes clear to students that in the traditional curriculum delivery, teachers of different content area subjects do not talk to each other, collaborate, or make connections for the students. STEAM changes that to show the students the way the real world works! And hopefully, to eliminate that age-old question, "Why do I have to know this?"

It is important for content area teachers to work together, and it is helpful to do that by establishing a theme that can overarch the learning in the five content areas. Usually, there is an open-ended question to frame the expected learning from each of the content areas. The goal is to remove the barriers between the various content areas and to guide the collaboration. Practical application in the real world makes this a requirement.

ANNOTATED PROFESSIONAL RESOURCES—STEM/STEAM

A great resource is http://www.educationcloset.com—a site with a lot of very useful information about STEAM provided by Susan Riley, an Arts Integration Specialist. Her books have been reviewed here, *No Permission Required: Bringing STEAM to Life in K-12 Schools* and *Steam Point: A Guide to Integrating Science, Technology, Engineering, the Arts and Math through Common Core*. Liz Knowles recently attended an online conference—all about STEAM Connectivity. Riley is most interested in integrating the arts and provides, on her website, free lesson ideas called "Lesson Seeds for Arts Integration." Riley sells excellent English Language Arts Integrated Curriculum Maps for K-5 from her website.

Another site that is an excellent resource for locating books about STEAM topics and information on STEAM authors is http://www.teachingbooks.net. Nick Glass does a great job of providing valuable materials for teachers and students about books, authors, and illustrators, and the subscription fee for access to the site is very reasonable for schools.

Bybee, Rodger W. *The Case for STEM Education: Challenges and Opportunities.* National Science Teachers Association, 2013.

> This book has a publication date of 2013, and it therefore describes the beginnings of the STEM program, which is important in order to understand the background for where we are with this educational movement today. It begins by describing the challenges for creating a STEM education. It describes how STEM education reform is different from other educational reforms and presents an opportunity for school systems to develop a coherent strategy for delivering a STEM education.

Johnson, Carla C., Erin E. Peters-Burton, and J. Moore. *STEM Road Map: A Framework for Integrated STEM Education.* Routledge, 2015.

> This resource has a publication date of 2015 and focuses solely on STEM. It is an edited compilation of the efforts of a number of authors. It begins by stating that a roadmap is necessary in order to plan a STEM program, and it discussed the history of its development. Emphasis is placed on the importance of the integration of the four different disciplines. The second part of the book is a STEM curriculum roadmap for grades K-12. The third part of the book is focused on resources for implementation and suggestions about effective assessment. It also discusses making STEM accessible to all learners and making sure that there is adequate professional development to provide faculty with the necessary skills to implement a top-notch STEM program.

Maslyk, Jacie. *STEAM Makers: Fostering Creativity and Innovation in the Elementary Classroom.* Corwin, 2016.

> This resource should be part of a STEAM professional collection because it is extremely thorough and helpful. It begins with the history of STEM and discusses the maker movement and project-based learning, as well as state testing, common core, and science standards. It is made clear that the support of school leadership, parents in the community, and teachers is an absolute necessity. There are cautionary statements about starting small and being aware of possible obstacles. It also discusses engineering and how it should be taught in the elementary grades. Makerspaces should be made available to all students, including English language learners, students with autism, students with speech and language difficulties, and so forth. Schools and their STEAM programs are highlighted. There are appendices with lots of very helpful information as well as a website http://www.steam-makers.com/.

Myers, Ann and Jill Berkowicz. *The STEM Shift: A Guide for School Leaders.* Corwin, 2015.

> Another resource about STEM that shares the reason STEM began and how important it is as part of the 21st century learning environment. STEM is a program for students of

all ages. It is important that we remember that graduates must compete globally and therefore they have to be prepared for the skills needed to be competitive. Students with disabilities, those living in poverty, and English language learners all should be part of a STEM program. The second part relates the necessity of leadership in developing a STEM program and how important it is to plan for all age levels. Professional development is extremely important as is changing schedules in order to accommodate a STEM program. The value of collaboration, trust, and community building when developing a STEM program cannot be overlooked.

Riley, Susan M. *No Permission Required: Bringing STEAM to Life in K-12 Schools.* Visionyst Press, 2014.
 Riley calls STEAM integration a disruption because focusing on arts integration, STEAM, project-based learning, and inquiry-driven instruction certainly does seem like a disruption to the traditional ways of education and the "that's the way we've always done it" mantra! The author shares strategies, behaviors, community, environments, and curricular development. The last part provides lessons and assessments for K-12 and references.

Riley, Susan M. *Steam Point: A Guide to Integrating Science, Technology, Engineering, the Arts and Math through Common Core.* Visionyst Press, 2012.
 Part one discusses curriculum mapping and the importance of building a frame for a STEAM program that includes the arts. Part two is lesson plans that can also be found on the education closet website (https://educationcloset.com/). Part three discusses assessments—the different types of assessments: integrated, formative, portfolio, and performance. There are appendices with additional materials and resources.

Sousa, David A. and Thomas Pilecki. *From STEM to STEAM: Using Brain-Compatible Strategies to Integrate the Arts.* Corwin, 2013.
 Even though this has a publication date of 2013, it is important because it provides brain-compatible strategies for transitioning from STEM to STEAM. It begins with a discussion of the power of the arts and how they are basic to the human experience. It discusses brain organization, thinking and learning, and divergent and convergent thinking. A program such as STEAM challenges the brain and actually changes the brain. The arts develop creativity—one of the 21st century skills. The book then addresses questions about how art can be integrated when the teacher is not an artist! STEAM is important in the entire K-12 program, and a variety of lesson suggestions are provided. Lastly, the significance of professional development to maintain a quality STEAM program is shared including the value of peer coaching, study groups, action research, and workshops for faculty.

Chapter 2

PROJECT-BASED LEARNING

THE CASE FOR PBL AND STEAM

- Real-world relevance and application
- Focus on the 21st century workplace skills
- Address standards and content requirements if well-planned
- Provide opportunities for students to find creative ways to use technology
- Connect students with the community
- Can give students a sense of purpose as they solve real-world problems
- Begin with a problem, question, challenge—brainstorming required
- Timeframe for work must be assigned by the teacher and a design plan created
- First there must be authentic research—online, interviewing experts, reading books and articles, watching informative videos, determining what has already been tried
- Research usually uncovers more questions—possibly the need for a narrower focus
- Keep a journal—steps taken, things tried, results, or going "back to the drawing board"
- Share results with peers and the community through effective presentation skills

Project-based learning is a natural fit with STEAM and the making process. Project-based learning requires driving questions or essential questions, to get to the heart of the problem, issue, or challenge. The inquiry process should have a timeframe—a reasonable amount of time should be designated to research, read, interview, and find out everything possible to create pathways to solutions. Next would be maker options to create a sample product or prototype guided by the engineering design process (ask, imagine, plan, create, improve).

It is important that students know that their solution attempts may not work and that they may have to go "back to the drawing board," and that is acceptable and encouraged. Students need to constantly reflect on what they are doing and evaluate their attempts and results. It is important to work together and value the suggestions of all team members.

The final part is sharing what has been learned and created with their peers and the community. The sharing portion provides practice for fine-tuning presentation skills—speaking before an audience, creating a clear and concise demonstration of the process, sharing prototypes, and answering questions—all valuable skills for future careers.

ANNOTATED JOURNAL ARTICLES

Gidcumb, Brianne. "PBL and STEAM: Do They Intersect?" *Education Closet*. October 29, 2016. https://educationcloset.com/2014/05/23/pbl-and-steam-where-do-they-intersect/
    The Education Closet (https://educationcloset.com/) provides an abundance of resources on STEAM, and this article presents a list of ways how STEAM and PBL work well together: both are all about process, they integrate multiple content areas—just like in real-world situations—and they can both be aligned with standards and content. Adding the arts enhances the process in a variety of ways. It is important for the students to own the problem, process, and the solution.

Miller, Andrew. "PBL and STEAM Education: A Natural Fit." *Edutopia*. May 20, 2014. https://www.edutopia.org/blog/pbl-and-steam-natural-fit-andrew-miller
    This article reiterates information that problem-solving, creativity, critical thinking, and collaboration are part of both PBL and STEAM. The author reminds us that one of the most important parts of this process is allowing the students to have voice and choice—to make the work their own.

"STEAM + Project-Based Learning: Real Solutions from Driving Questions." *Edutopia*. January 26, 2016. https://www.edutopia.org/practice/steam-project-based-learning-real -solutions-driving-questions
    The Charles R. Drew Charter School in Atlanta, Georgia, is a STEAM school that uses PBL to provide the structure for learning. The elementary grades provide multiple PBL units during the school year following the resources delivered online by the Buck Institute of Education (www.bie.org). This resource is especially helpful in preparing units according to standards, and it helps with ideas for driving questions. Even though the teachers help to prepare the units, they all know the value of problem-solving locally and having the students make the final decisions.

Jolly, Anne. "The Search for Real-World STEM Problems." Education Week Teacher. July 21, 2017. https://www.edweek.org/tm/articles/2017/07/17/the-search-for-real-world -stem-problems.html
    It is a challenge to settle on a project that meets all criteria. It has to be a real issue that the students can relate to—in their school or community or even around the world. Then it cannot be hugely out of the realm of possibilities—it has to be something that they can actually handle. Ideally, it should have more than one possible solution. Then it has to be aligned with standards and content curriculum. And lastly, it should be set up to

follow the process that the students have been taught—hopefully the engineering design process. Despite all these parameters, it can be an exciting and fulfilling challenge for the students with amazing results!

Markham, Thom. "STEM, STEAM, and PBL." ASCD Edge. 2012. http://edge.ascd.org/blogpost/stem-steam-and-pbl
> Another great resource for PBL is http://pblglobal.com. Markham is a PBL consultant and the author of a PBL design and coaching guide. His website has a number of useful tools for the PBL classroom. PBL is a very natural fit for STEAM programs. PBL is a process that is inquiry-based. The focus in any STEAM program should not be the technology. Technology is a tool—it helps with problem-solving and design creation. The students still need to do the detective work and demonstrate perseverance. Attention must be given to the standards and content curriculum. Teamwork is an important part of this process—it is important to have everyone working together to solve the problem and create a viable and practical solution.

Larmer, John and John Mergendoller. "Seven Essentials for Project-Based Learning." *Educational Leadership* 68:1 (2010): 34–37.
> The authors share seven essentials for PBL. Both are directors at the Buck Institute for Education (www.bie.com)—an organization that provides valuable resources on project-based learning. First, there must be a solid reason for the project selection—it has to be real and meaningful. All projects should be framed and driven by an overarching essential or driving question that is open-ended. Students have to be involved in the process, which includes the actual selection of the project. Ideally, the 4C's should form the framework—collaboration, communication, critical thinking, and creativity. Students should be encouraged and inspired to think outside of the box, and all ideas should be addressed as worthy by the team. The students need to know and feel comfortable with the concept of failures being an important snag and actually learning experience along the path to a solution/end product. And lastly, students should be given an opportunity to share their results with their peers and the community.

Gorman, Michael. "Part 1: STEM, STEAM, Makers: Connecting Project Based Learning (PBL)." 21st Century Educational Technology and Learning. December 22, 2016. https://21centuryedtech.wordpress.com/2016/07/05/part-1-stem-steam-makers-connecting-project-based-learning-pbl/
> Michael Gorman (http://www.21centuryedtech.wordpress.com) offers a wide variety of information about the integration of technology and STEM/STEAM/PBL on his website. Each content area provides a different role in the education process. STEAM provides the much-needed integration of content areas. PBL focuses on the process for delivery of content. Technology provides the means and tools for increased productivity.

Larmer, John. "It's a Project-Based World." *Educational Leadership* 73:6 (2016): 66–70.
> If students are given the opportunity to learn using PBL, they can see that their work is important and what they do can actually make a difference—locally or even globally. The same steps as seen in prior articles and sources are enumerated here: selecting a worthy problem with student input, working through a design process, and beginning with serious and rigorous research. The importance of perseverance is key, and revision along the way is required. The end result should be shared with peers and more broadly with the community.

Waters, Patrick. "Project-Based Learning through a Maker's Lens." *Edutopia.* July 9, 2014. https://www.edutopia.org/blog/pbl-through-a-makers-lens-patrick-waters
> The word "learner" is synonymous with the word maker (Waters 2014). This article reviews similar steps in the process of connecting the PBL instructional delivery method with making. It is necessary to connect the project with standards, which is usually quite easy to do. Ideally, a project should connect as many of the content areas as possible and should begin with an essential question. A design process should follow. Waters cautions teachers regarding focusing on the process—not the product and remembering that the teacher does not always have the answer. Another caution is to make sure that the project is not too large or too impossible for the students tackling it.

Miller, Andrew. "In Search of the Driving Question." *Edutopia.* August 30, 2017. https://www.edutopia.org/article/search-driving-question
> What makes a good driving question is discussed in this article. An essential question, as part of the design process, focuses on the desired outcomes. A driving question can be varied—it can be debatable or results oriented. It can also be role-oriented where the students are asked to take on certain roles as they work on the solution/product. Teachers can create a driving question focused on an appropriate verb/action. A consideration in developing a driving question would be the age of the students. And a teacher could ask the students to help in fine-tuning a driving question for a project.

Duke, Nell K. and Anne-Lise Halvorsen. "New Study Shows the Impact of PBL on Student Achievement." *Edutopia.* June 20, 2017. https://www.edutopia.org/article/new-study-shows-impact-pbl-student-achievement-nell-duke-anne-lise-halvorsen
> The results of this PBL study show that PBL is having a positive impact on high-poverty, low-performing 2nd grade students. The factors that made the difference were teacher encouragement and alignment with standards. PBL as the instructional delivery method required teachers to utilize research-supported techniques. The study will continue and expand in the future.

ANNOTATED PROFESSIONAL RESOURCES—PROJECT-BASED LEARNING

Buck Institute for Education is the go-to location for information and resources on project-based learning (http://www.bie.com).

Bender, William. *Project-Based Learning: Differentiating Instruction for the 21st Century.* Corwin, 2012.
> This book begins with the basics on project-based learning and has some information about research, which is interesting even though it is an older publication date. There is information about using PBL as a schoolwide means of instructional delivery. It includes material about project design and effective use of instructional technology to support project-based learning. There is a very helpful chapter on assessment options.

Boss, Suzie. *Implementing Project-Based Learning.* Solution Tree, 2015.
> Boss is one of the better-known authors in the area of PBL. She first shares PBL basics and states the fact that the roles have changed with PBL—the emphasis falls more on the performance of the students than on the teacher. There is a chapter on making global

connections with "geoliteracy" projects. Boss devotes chapters on several different types of literacy projects. She ends with a discussion of the challenges of using PBL, which includes content coverage and assessment.

Cooper, Ross and Erin Murphy. *Hacking Project Based Learning: 10 Easy Steps to PBL and Inquiry in the Classroom*. Times 10 Publications, 2016.
The 10 hacks are promote risk-taking, collaboration skills, find worthy content, create a vision, use inquiry, ownership of assessment, provide abundant feedback, use mini-lessons, make sure there is understanding, and make a grand ending!

Elliott, Lori. *Project Based Learning for Real Kids and Real Teachers*. SDE, 2016.
This resource is divided into three sections: the first provides background information on PBL basics; the second discusses the design challenge, innovative inventions, and getting the entire school on board; and the last part focuses on the importance of planning.

Farber, Katy. *Real and Relevant: A Guide for Service and Project-Based Learning*. Rowman & Littlefield, 2017.
This resource discusses service learning and how project-based learning and service learning can be connected. How does one build a service and project learning climate in a school? The book goes through ideas, planning, and partnering with the community while working together in teams. It also shares information on assessments and reflections and what technology tools can support this kind of joint venture between service learning and project-based learning. It ends with a chapter on makerspaces and genius hours and provides interviews with teachers who are currently making this connection.

Fehrenbacher, Tom and Randy Scherer. *Hands and Minds: A Guide to Project-Based Learning for Teachers by Teachers*. CreateSpace Independent Publishing Platform, 2017.
This resource starts with a chapter on finding project ideas and then designing the project including essential questions, setting goals, and developing a process. It also discusses the importance of sharing findings and presenting results. It emphasizes the value of community relationships. It also discusses assessment, portfolios, and all kinds of reflections.

Kraus, Jane and Suzie Boss. *Thinking through Project-Based Learning: Guiding Deeper Inquiry*. Corwin, 2013.
This PBL information is framed into sections. The first describes PBL in detail as to setting things up and creating the atmosphere for project-based learning to thrive. A second section talks about different disciplines and how they can be intertwined through project-based learning.

Larmer, John, John R. Mergendoller, and Suzie Boss. *Setting the Standard for Project Based Learning: A Proven Approach to Rigorous Classroom Instruction*. ASCD, 2015.
Larmer and Mergendoller are executives with the Buck Institute for Education, which is all about PBL. This resource provides everything you need to know about PBL, including information on what research says about the effect of project-based learning. It also tells about designing, managing, and leading the PBL effort. It gives advice for PBL summer programs as well.

Laur, Dayna and Jill Ackers. *Developing Natural Curiosity through Project-Based Learning: Five Strategies for the PreK–3 Classroom*. Routledge, 2017.

Developing ideas for projects and learning experiences is the first subject in this book. There are five strategies discussed in this book, and the first one is mapping standards to the project. The remaining four strategies provide guidance for implementing a PBL program. The last chapter discusses the importance of the 21st century skills focus: communication, critical thinking, collaboration, and creativity.

McDowell, Michael P. *Rigorous PBL by Design: Three Shifts for Developing Confident and Competent Learners.* Corwin, 2017.
The three shifts are clarity, challenge, and culture. This resource has a long list of online resources. It begins with the usual discussion of PBL and seems to be focused mainly on high school and middle school students. An important part of the content is designing PBL for students' confidence and competence so that they understand expectations from the beginning. It also tells about structuring the whole PBL experience and establishing a culture among the students to make sure that they are comfortable with the process.

Warren, Acacia M. *Project-Based Learning across the Disciplines: Plan, Manage, and Assess through +1 Pedagogy.* Corwin, 2016.
Under the general category of planning, there is information about PBL and how important it is to plan with the end in mind. There is a chapter on making a connection with the standards from all disciplines and a chapter on essential questions and how to develop them. The next general category is managing and the importance of inquiry and research. Another chapter talks about how to utilize technology in the most efficient way. The last general category is assessment and the different ways to assess PBL—especially focusing on the 21st century skills.

Chapter 3

MAKERSPACES

THE CASE FOR MAKERSPACES

- Makerspaces are important because activities are hands-on, multisensory, multidisciplinary, and concrete.
- Students focus on the process not the product.
- Maker education is self differentiating—age levels and genders are blurred.
- Activities are multidisciplinary and authentic.
- A maker education activity can include science, technology, engineering, arts, and math with ease.
- A byproduct of maker education is resilience—if something does not work or things do not go as planned, the team of students works together to solve problems and bounce back from disappointments.
- A makerspace can be anywhere: it can be a center in a classroom, on a mobile cart, in a former computer lab, or in a library.
- Projects most often are collaborative efforts that include creativity, communication, and critical thinking.
- Most important thing to learn from the maker movement is that it is the doing that is what matters.
- Asking questions and asking for help is the norm, and peer tutoring and collaboration are encouraged.

The maker movement has its root in the visions of Papert and Piaget and others, who championed constructivism and constructionism. It is in many ways contradictory to curriculum guides, pacing charts, and high-stakes testing. But there are ways to tie it to standards and the instructional method of project-based learning. It is driven by essential, driving questions and encourages students to collaborate and search for solutions. It promotes perseverance and is the answer to that age-old question, "Why do I need to learn this?"

A makerspace can be anywhere and does not really need expensive equipment. It promotes service learning, investigations, and reflection. It teaches students that there is more than one correct or possible answer to a problem, and failure is allowed and encouraged because it promotes deep thinking and resilience. It also provides students with practice for real-world application.

ANNOTATED JOURNAL ARTICLES

Meyer, Leila. "7 Tips for Planning a Makerspace." *The Journal*. February 23, 2017. https://thejournal.com/articles/2017/02/23/7-tips-for-planning-a-makerspace.aspx

The author describes seven tips for planning makerspaces. Many are common knowledge and shared in numerous other articles, but one is an especially good idea. It is important to search for established makerspaces to visit. They can be in schools, libraries, and even private businesses. By seeing what works and being able to ask questions, stakeholders and planners can then feel more confident about planning a makerspace of their own. Stakeholders should include school officials, teachers, parents, community leaders, and, most important, students! The final advice is to start small.

Anglin, Nick. "A Student Maker and the Birth of a Startup." *Edutopia*. August 27, 2015. https://www.edutopia.org/blog/student-maker-birth-of-startup-nick-anglin

This is an important piece because it is the story of a frustrated and school-hating student. Because of poor grades, he had to attend a summer class, which was a summer maker camp. The camp challenge was to create a project related to something the student loved and incorporate some type of technology and possibly start a business. Nick, an avid baseball fan and a catcher and pitcher, decided on creating a practice strike zone for pitchers. He went through many iterations of his design and prototype and finally entered a contest and was named the Student Entrepreneur of the Year by the Business Innovation Council in his town. He learned coding, electronics, woodworking, patent searches, communication, and business skills. The entire experience gave him a reason to go to school.

Stager, Gary S. "Unconventional Wisdom about the Maker Movement." Invent to Learn. Winter, 2017. https://inventtolearn.com/unconventional-wisdom-about-the-maker-movement/

The author, who is one of the premier leaders in maker education, shares many ideas in this report. One that bears repeating is that in the word "makerspace"—maker is the most important part—the actual space is inconsequential. Size does NOT matter! He also qualifies 3D printing as a non-requirement. While computer programing is not required, it is indeed important because students are going to be looking for jobs in a technically sophisticated world. Also—a makerspace can be created on a shoestring—it is the meaningful problem-solving and the construction of knowledge that are the focus. Also, time is important, and meaningful making probably cannot be accomplished in a 42-minute class period!

Herold, Benjamin. "Researchers Probe Equity, Design Principles in Maker Ed." *Education Week* 35 (2016): 8–9.

Herold shares that maker education provides hands-on activities supporting academic learning and the development of a mindset that encourages experimentation, growth, collaboration, and community. The goal is to solve a problem by creating a prototype and then sharing the results with an audience. But remember, the process is more important than the results.

Jarrett, Kevin. "Digital Shop Class: Fun and Profitable." *Edutopia.* January 5, 2017. https://www.edutopia.org/article/digital-shop-class-fun-profitable-kevin-jarrett
  An entrepreneurial twist to making is shared by Jarrett who has created a digital shop class with a store! He has turned the makerspace activities into students designing and producing gifts and keepsakes. Students have learned design and production as well as marketing and all the skills required for running a business. They developed an online store and also worked on a freelance basis to sell their designs. Their initiatives were tied directly to community needs and solutions to problems and challenges that were affecting the entire town.

Busch, Laura. "How Should We Measure the Impact of Makerspaces?" *EdSurge.* May 31, 2017. https://www.edsurge.com/news/2017-01-09-how-should-we-measure-the-impact -of-makerspaces
  There is really no right way when it comes to using makerspaces—standardized test scores may never accurately show the impact that makerspaces have on student achievement. However, it is the combination of design thinking, service learning, and 21st century skills that are changing student learning. The future jobs for these students require skills that are part of makerspace education. Some of those are problem-solving, critical thinking, creativity, working together, decision making, negotiation, and flexibility.

Wolfe, Aiden. "How to Turn Any Classroom into a Makerspace." *Edudemic.* April 9, 2015. http://www.edudemic.com/turn-classroom-makerspace/
  The author declares that everybody likes to improve the world around them, and the maker movement solves that need with problem-solving, innovative design, and construction. He shares information about hydroponic gardening and the website (https://kidsgardening.org/) that shares classroom and school gardening tips. Another connected site is https://www.3dponics.com/ that provides design files for creation and customization of class garden setups. Wolfe also shares a site for those who have 3D printers: https://www.yobi3d.com/, which is a search engine for finding design plans for toys. Another site he recommends is http://appinventor.mit.edu/explore/ for inventing apps.

Fleming, Laura and Ross Cooper. "Makerspace Stories and Social Media: Leveraging the Learning." *Edutopia.* September 1, 2016. https://www.edutopia.org/blog/makerspace -social-media-leveraging-learning-ross-cooper-laura-fleming
  The authors suggest that students should make a one-minute video of the results of their makerspace project. The video can be archived or shared on social media. That exercise can also result in a lesson on appropriate feedback and comments!

Dillon, P. Mathew. "Makerspace Technology: Is It Right for Your School?" *Edutopia.* January 31, 2017. https://www.edutopia.org/discussion/makerspace-technology-it-right -your-school
  Dillon says it is important to impress upon stakeholders that makerspace technology is not merely a fad, but it is actually deeply rooted in educational history with constructivist theories inspired by others. They are considered the chief theorists among the cognitive constructivists, but all this has been lost because of the pressure to teach to the test.

Fleming, Laura and Billy Krakower. "Makerspaces and Equal Access to Learning." *Edutopia*. July 19, 2016. https://www.edutopia.org/blog/makerspaces-equal-access-to-learning-laura-fleming-billy-krakower

The authors state that the positives of the makerspace movement include access by all learners, the creation of a community of learners, and the students being allowed to make plans, decisions, and mistakes pretty much on their own.

O'Brien, Chris. "Makerspaces Lead to School and Community Successes." *Edutopia*. March 21, 2016. https://www.edutopia.org/blog/makerspaces-school-and-community-successes-chris-obrien

O'Brien calls the makerspace movement the new industrial revolution! Makerspace education helps students develop grit and the metacognitive skills needed to review what they have done and get past any failures. Teachers are responsible for making the connections between the content curriculum, standards, and the makerspace activities. Makerspaces work much better as tools to motivate students than any pizza party!

Bentley, Kipp. "Makerspaces: New Prospects for Hands-On Learning in Schools." *Center for Digital Education & Converge Magazine*. February 12, 2017. http://www.centerdigitaled.com/blog/makerspaces.html

Bentley states that makerspaces provide students with tools and materials to do hands-on projects. Often if teachers are hesitant about using makerspace technology, all they need to do is leave it to the students to work out, and they all become experts on using these tools and then will teach others.

Martinez, Sylvia. "Making for All: How to Build an Inclusive Makerspace." *EdSurge*. May 10, 2015. https://www.edsurge.com/news/2015-05-10-making-for-all-how-to-build-an-inclusive-makerspace

Martinez says we ring a bell every 42 minutes to move students quickly to a new subject, and then we are upset with their short attention spans! The makerspace movement should not be thought of as an alternative way to learn but what learning should be. Maker education is an ongoing process of finding interesting challenges for students and helping them build skills needed to complete them. Also, it is important to be sensitive to the gender of students and make all students comfortable and valued.

Ullman, Ellen. "Making the Grade: How Schools Are Creating and Using Makerspaces." *Tech Learning*. March 24, 2016. http://www.techlearning.com/resources/0003/making-the-grade-how-schools-are-creating-and-using-makerspaces/69967

A mobile solution is perfect for schools that lack physical space according to the author (Ullman 2016). She suggests a creativity cart, a robotics cart with iPads and an Osmo, a math and science cart, and any others that can be imagined.

Spencer, John. "Why Every Classroom Should Be a Makerspace." July 19, 2017. http://www.spencerauthor.com/classroom-makerspace/

The skills that students are learning in makerspaces, according to Spencer (2017), are valuable skills for success in life. He thinks that every class should be a makerspace. This is not an easy job—to transform spaces held hostage by curriculum maps, standards, and test prep and turn them into areas of student-driven problem-solving and creativity. Maker style learning does not have to be limited to STEM/STEAM classes—it can be part of English language arts and social studies, too.

Fleming, Laura. "Out of the Box Approach to Planning Makerspaces." Worlds of Learning. April 4, 2017. https://worlds-of-learning.com/2017/04/04/box-approach-makerspaces/
A large part of her second book, Fleming (2017) says, was highlighting and showcasing great makerspaces. While many are similar, some are very different, and here she shares specifics about some of them. One in particular was a community care makerspace for nurses and health professionals where various new tools and changes in old ones were created to solve problems and make life easier and more comfortable for patients and nurses alike.

Harper, Charlie. "The STEAM-Powered Classroom." *Educational Leadership* 75 (2017): 70–74.
A call to action for any teacher leader who dreams of transforming learning at school is described by the author (Harper 2017). A challenge should require students to use inquiry, critical thinking, and problem-solving skills while the teacher moves between groups to ask questions, evaluate progress, and introduce appropriate vocabulary. Start with essential, driving questions and utilize backwards design and planning.

Ullman, Ellen. "How Schools Make 'Making' Meaningful." Tech Learning. September, 2017. http://www.techlearning.com/resources/0003/how-schools-make-making-meaningful/70727
Maker education should allow students to connect to their curriculum, their community, and the world according to Ullman (2017). By my estimation, the best project she shared was reading *The True Confessions of Charlotte Doyle*. Students were asked to create a vending machine filled with all sorts of creative items that might have been helpful to Charlotte as she endured the long and very difficult sea voyage from England to the United States in the 1800s.

Wilkinson, Karen, Bevan Bronwyn, and Mike Petrich. "Tinkering Is Serious Play." *Educational Leadership* 72:4 (2014–2015): 28–33.
The three authors of this article state that scientific principles are a large part of maker activities. Questioning, defining problems, testing solutions, responding to feedback, and generating explanations are all important parts. Almost no directions and certainly no model of the outcome are to be presented by the teacher. Students need to go it alone—relying on their collective problem-solving skills. There is no single correct answer, and failure and frustration are to be endured for the greater good!

ANNOTATED PROFESSIONAL RESOURCES—MAKERSPACES

Anderson, Chris. *Makers: The New Industrial Revolution*. Crown Business, 2014.
The new industrial revolution is all about inventions and the future and the fact that thanks to makerspaces and the technology—we are all designers now! There is a discussion of desktop factories, open hardware, and financing the maker movement. The epilogue is entitled "The New Shape of the Industrial World." The appendix includes some very helpful information about CAD, 3D printing and scanning, laser cutting, CNC machines, and electronics.

Clapp, Edward P., Jessica Ross, Jennifer O. Ryan, and Shari Tishman. *Maker-Centered Learning: Empowering Young People to Shape Their Worlds*. Jossey-Bass, 2016.
This book begins with an overview of the benefits of maker-centered learning and describes teaching and learning there. Maker education provides a huge sense of

empowerment that is facilitated by design elements. The author describes maker-centered teaching and learning in action currently and shares thoughts about the future of maker education. There is an appendix with resources.

"Classroom Playbook." *Maker Faire*. Maker Media, May 2016.
This guide was written to help a teacher get the best experience for the students attending a maker faire. There is the definition of a maker, including a list of things that a maker believes. There is information about making your own sketchbook and a list of educational goals to review before the visit. There are suggested on-site activities and more about a life-size mouse trap, which is a favorite maker Rube Goldberg machine. There are things to do for recapping and reviewing after the visit and suggestions about starting a maker club or makerspace at school.

Dougherty, Dale. *Free to Make: How the Maker Movement Is Changing Our Schools, Our Jobs, and Our Minds*. North Atlantic Books, 2016.
Dale Dougherty, creator of *MAKE:* Magazine and the Maker Faire, provides in this book chapters describing how we are all makers of one kind or another: amateurs, enthusiasts, and professionals. Interaction and innovation in the community—peers, local, and global—are very valuable. Finally, he defines making as learning, working, caring, and the future. The appendix includes an interesting chronology of the maker movement beginning with the year 2000.

Fleming, Laura. *The Kickstart Guide to Making Great Makerspaces*. Corwin, 2017.
This is the recent second book from Laura Fleming, the librarian and champion of the makerspace movement. From the table of contents, the first chapter lists seven attributes of a great makerspace: personalized, deep, empowering, equitable, differentiated, intentional, and inspiring. The next chapter discusses making great makerspaces and the admonition that no cookie cutter makerspaces are allowed. The chapter also talks about creating conditions to inspire and then assessing creativity without squashing it.

Fleming, Laura. *Worlds of Making: Best Practices for Establishing a Makerspace for Your School*. Corwin, 2015.
Laura Fleming is one of the first librarians thoroughly involved in the maker movement. This book gives a history of the maker movement. It then tells how to plan and set up a makerspace and work to create a maker culture in school. The next part discusses makerspaces and standards. The book continues to share information about makerspaces and the school library and makerspaces as a unique learning environment. The final chapter discusses showcasing student creations, which is a very important part of the maker movement.

Lang, David. *A Beginner's Guide to the Skills, Tools, and Ideas of the Maker Movement*. Maker Media, 2014.
This book is the second edition of a book that was originally written four years ago. The author said the changes in the four years have been in tools and resources for makers. This book begins a lengthy description on the maker mentality and then provides information on access to tools and craftsmanship. There is important detail about digital fabrication and makers going pro and actually making something that matters.

"Maker Club Playbook." Young Makers. Maker Ed, January 2012.
   The Maker Club Playbook begins with an explanation of the maker movement and how it all began. It further explains the potential impact on education. There are descriptions of a variety of maker projects with lots of great information. There are tips on how to be a successful mentor. And there is also a whole chapter on starting a maker club. Guidance about going to a maker faire and what to look for and then what to do when you return is included. There are even some snapshots of current schools with Maker Clubs. A chapter on resources, including a sample proposal and budget to submit to a funder for setting up a makerspace, ends this very useful guide.

"Makerspace Playbook School Edition." *Makerspace.* Maker Media, Spring 2013.
   Makerspace Playbook School Edition is a very valuable resource even though it has a publication date of 2013, and it still has a huge amount of important information. There is background about the importance of the maker movement and maker education. It describes the process of finding a physical place for a makerspace and reports on tools and materials, providing a very complete list. There are facts on safety and detailed descriptions of the responsibilities of various participants—including the manager, coach, research librarian, mentors, and the students. There is a Maker Manifesto, which is an excellent resource. There is a timeline for how to plan a year of making, and there are many suggestions for projects ideas. A very complete resource section can be found at the end.

Martinez, Sylvia Libow, and Gary S. Stager. *Invent to Learn: Making, Tinkering, and Engineering in the Classroom.* Constructing Modern Knowledge Press, 2013.
   This book is probably one of the most important books for detailing making in the classroom. The table of contents shows that the book begins with the history of the maker movement and the connection to constructivism. It also tells about the importance of design models and lists the elements of a good project (there are eight). The next chapters tell about teaching—less "us" and more "them" and the changes in the learning environment. There is also a great resource chapter.

Peppler, Kylie A., Erica Halverson, and Yasmin B. Kafai. *Makeology: Makerspaces as Learning Environments.* Routledge, 2016.
   The book has a foreword by Dale Dougherty. There are three editors for this book, so chapters are written by different contributors. The introduction is about learning to make in museums, designing for resourcefulness, and information on a community-based makerspace. There are additional chapters about family creative learning and play-shop with preschoolers. The authors describe bringing making to schools and classrooms, higher education, and there is a whole section about online makerspaces. Other topics include youth, tinkering online, MOOCs, and social media. The role of communication in the maker community, design thinking, collaboration and community building, and building process-oriented documentation complete the topics discussed in the book.

Spencer, John. *Launch: Using Design Thinking to Boost Creativity and Bring Out the Maker in Every Student.* Dave Burgess Consulting, Inc., 2016.
   Spencer begins by highlighting the importance of creative classrooms and the creative approach. He provides a description of the "launch cycle" and look, listen, and learn. It is important to ask tons of questions to understand the information and navigate the details. Improving the product is required before it is time to launch. The book ends with frequently asked questions, lesson plans, and a notebook about creating a roller coaster.

Chapter 4

3D PRINTING

THE CASE FOR 3D PRINTERS

- There are major applications in medicine and healthcare.
- The price for 3D printers is dropping.
- Manufacturers are creating curriculum guides to accompany their 3D printers.
- The 3D printer business is over $30 billion.
- 3D printers are beginning to use different materials other than resin and plastics.
- Design software programs can be free and easy to use.
- Students in kindergarten can successfully design prototypes and use 3D printers.
- Many new 3D printers are PC compatible.
- Unlike traditional mass manufacturing, with 3D printing there is no waste.

3D printing is now a booming business. It is forecast to soon become as popular as the personal computer. Originally, a 3D printer was extremely expensive, but now the price has come down dramatically; and for both education and personal use, there is a wide variety of options.

The current 3D printer is easy to use, and there is free software that is also available that helps the student or amateur to create just about anything. The latest 3D printers for use in schools now come with complete curriculum so that the implementation into a maker program education program is seamless.

On the other hand, maker education can still be thriving without a 3D printer—it is not an absolute requirement. The benefits, though, are important along with the technology that goes with it because those skills are currently in high demand in the workplace and will be almost a requirement in the very near future.

ANNOTATED JOURNAL ARTICLES

Carbon, Kim. "Going from STEM to STEAM, Transforming Education with 3D Printing." *364-Type-A Machines.* April 21, 2017. https://www.typeamachines.com/blog/stem-to-steam-transforming-education-with-3d-printing

> Almost all the articles about 3D printing discuss the value of this "science" for our students' future careers. Carbon is well aware of the skills that will be needed and the application and practice of skills such as problem-solving and critical thinking as part of the design process for 3D printing. The opportunities to create real-world models to better understand whether the plan will work or not also allow for students to see the value of mistakes and then the importance of corrections and adjustments. This is the engineering design process in action.

Thornburg, David D. "The 3D Printing Revolution in Education." *ESchool News.* January 29, 2016. https://www.eschoolnews.com/whitepapers/the-3d-printing-revolution-in-education/

> In the scheme of things regarding student engagement with learning, Thornburg is one of the leading experts. He understands the constructivist philosophy of Papert from MIT and therefore the value of hands-on learning for today's students. Maker education and 3D printing can be tied to standards, and the benefit of being able to see ideas and creations come to fruition with a 3D printed object is very valuable. He feels that as the computer has become an everyday tool accessible to almost everyone, that is also the future for the 3D printer!

Adams, Caralee. "How to Use a Desktop 3D Printer in School." *WeAreTeachers.* July 17, 2017. https://www.weareteachers.com/the-non-tech-teacher-s-guide-to-using-a-3d-printer/

> According to the author, 3D printing is accomplished layer by layer from a single design file. The filament comes in plastic and resin. Plastic is recommended for school use because resin can give off strong fumes. There are many websites that offer digital designs for 3D printing. The pricing of printers has come down since they were first introduced, and a printer that allows for student viewing of the printing process is recommended. Constructing a prototype as part of a problem-solving project can be accomplished in any class; it does not have to be science or math!

Kane, Karen. "9 Ways Teachers Can Use a 3D Printer to Teach Math and Science." *WeAreTeachers.* February 24, 2017. https://www.weareteachers.com/3d-printing-math-science/

> The author gives concrete suggestions for using student-created 3D models in any class to aid in problem-solving. Many ideas and resources are available online for both teachers and students. The sites are often searchable if you are looking to connect a standard.

"3D Printing Educator Spotlight On: Jayda Pugliese, 5th Grade Teacher with a STEAM-Focused Classroom, Philadelphia." *3DPrint.com.* July 07, 2017. https://www.3dprint.com/180018/educator-spotlight-jayda-pugliese/

> This teacher has a fully-equipped STEAM classroom from writing grants and campaigning for donations. She learned how to do everything by watching YouTube videos and found the technology accessible and user-friendly. Her students have access to computers

and a wide variety of tools, supplies, and kits, and they know how to take care of the 3D printers and help others with the processes. Her students are getting hands-on experience designing objects to solve problems in their community.

"3D Printing Educator Spotlight On: Megan Finesilver, 2nd Grade Teacher, Speaker and Curriculum Developer, Colorado." *3DPrint.com*. July 17, 2017. https://www.3dprint.com/181092/spotlight-megan-finesilver/
> Here is an example of successful use of a 3D printer in a 2nd grade class. The teacher devised a plan to incorporate the 3D printer in her everyday lessons. She suggests using SketchUp and TinkerCAD because they are the best and FREE. The teacher feels that CAD skills will be a must for job seekers in the very near future.

Finley, Todd. "Jaw-Dropping Classroom 3D Printer Creations." *Edutopia*. June 30, 2015. https://www.edutopia.org/blog/jaw-dropping-classroom-3d-printer-todd-finley
> This article tells about 3D printing in high school, but the teacher also includes his two young sons who spend a lot of time with him in his classroom with the 3D printer. The teacher identifies good 3D template sites, such as Instructables, YouMagine, and Thingverse, and lists other helpful sites throughout the article. His students use the 3D printer in many different ways, and he says six-year-olds can design and make their own toys!

Mersand, Shannon. "What's New in 3D Printing." *Tech Learning*. September 2017. http://www.techlearning.com/resources/0003/whats-new-in-3d-printing/70729
> This is another article about high school 3D printing, but this one shares an important example of a senior design challenge that involved creating a prosthetic hand for a middle school student with cerebral palsy. Mersand also states that 3D printing companies are now including curriculum, lesson plans, and ideas for integrating the 3D printer into maker education.

Slavin, Tim. "What Is 3D Printing? Kids, Code, and Computer Science." *BEANZ*. January 31, 2015. https://www.kidscodecs.com/what-is-3d-printing/
> Most of the articles contain definitions of the various processes involved in maker education. Slavin adds a concise definition for 3D printing here. This includes the layering process and the length of time because that is significant. There is great importance in seeing the process, and he states that just about any age student can successfully use a 3D printer. This article provides a very complete glossary of 3D printing terms. The article was found in an online magazine called *BEANZ, The Magazine for Kids, Code, and Computer Science* by Owl Hill Media, which has some really good articles about technology for kids.

Tahnk, Jeana Lee. "6 Awesome Ways to Bring Your Kids' Ideas to Life with 3D Printing." *Mashable*. September 19, 2015. http://mashable.com/2015/09/19/kids-toys-3d-printing/
> Originally, it was hoped that 3D printing would take up the slack in manufacturing, but that has not happened yet because printing is singular and slow. However, it has become a booming business for common use. The global market for 3D printers has grown to almost $30 billion. The article includes some websites, but the most interesting thing was The Micro, which is a 3D printing venture that got began on KickStarter—and it is a portable, lightweight 3D printer that is now $299!

Wasserman, Todd. "Intersection of Creativity, Technology & Learning: A Conversation." *Worlds of Learning*. October 26, 2017. https://worlds-of-learning.com/2017/10/26/intersection-creativity-technology-learning-conversation/

> Wasserman says that maker education and 3D printing are part of the new industrial revolution. As the days go by, the technology gets better and cheaper and is therefore available to more people who then can problem-solve and create. The industry is booming—auto manufacturers are now using additive manufacturing (another name for 3D printing) to create prototypes and replacement parts. The 3D printing process is still too slow for mass production. The healthcare industry is using 3D printing for human "replacement parts." Wasserman predicts big changes in industry as 3D printing becomes more sophisticated.

Murphy, Kate. "UC Students Build 3D-Printed Prosthetic Hands for Kids." *USA Today*. February 16, 2017. https://www.usatoday.com/story/tech/nation-now/2017/02/17/students-build-3-d-printed-prosthetic-hands-kids/98071192/

> A group of biomedical engineering students at the University of Cincinnati have begun producing prosthetic hands in their lab for a small portion of the going rate and in a fraction of the usual amount of time. They have accomplished this with 3D printers. They are looking forward to the day when the 3D printer becomes more sophisticated, and they will be able to do more to help patients.

Hale, Brent. "175 Amazing Ways 3D Printing Is Changing the World." *3D Forged*. March 27, 2017. https://3dforged.com/3d-printing/

> This is an amazing and comprehensive list of links to information regarding things created by 3D printers. 3D printing is solving a host of worldwide problems that were, before, not solvable. And the best part is 3D printing is just starting. The report is divided into sections: environment and sustainability; entertainment and recreational use; science, research, and education; culinary arts; aerospace, aviation, and automotive; government and military; general manufacturing; fashion and apparel; medicine and healthcare; humanitarian efforts; and housing and home decor. Within each section there are 10 to 15 links. It is truly amazing and worth the effort to look through and share.

ANNOTATED PROFESSIONAL RESOURCES—3D PRINTING

Thornburg, David D., Norma Thornburg, and Sara Armstrong. *The Invent to Learn Guide to 3D Printing in the Classroom: Recipes for Success*. Constructing Modern Knowledge Press, 2014.

> Invent to Learn is one of the primary resources for the maker movement—so this is exciting addition to the resources they offer. The beginning shares information about 3D printing in general—challenges, integrating standards, how 3D printers work, and how students create designs. The rest of the book shares directions for a variety of projects. The book ends with information on finding more resources and materials.

Cano, Lesley M. *3D Printing: A Powerful New Curriculum Tool for Your School Library*. Libraries Unlimited, 2015.

> From the table of contents, it is easy to see that this is a beginner's guide to all things about 3D printing. The first section is all about the basics from selecting one to using it. Next comes a description of the 3D printer's role in education. Programming and applications are described, and then specific modeling programs such as TinkerCAD,

123D Design, 3DTin, and SketchUp are shared in more detail. There is information about modeling apps. There are several chapters describing the use of the 3D printer in math, science, social studies, fine arts, and language arts. The last chapter puts it all together, and there is an index of resources at the end.

France, Anna Kaziunas and Brian Jepson. *Make: 3D Printing: The Essential Guide to 3D Printers*. Maker Media, 2014.
This is another guide for beginners that includes getting started and a chapter that reviews some of the specific brands of 3D printers available. There is a section on all the different software programs and one on design for the beginners and getting started with Slic3r. There are chapters on scanning, printing your head in 3D, plastics and industrial materials, and service providers. Then at the end, there is information about dying and post processing your prints. And the last chapters have information about printing a humanoid and a white chocolate skull!

Kloski, Liza and Nick Kloski. *Make: Getting Started With 3d Printing: A Hands-On Guide to the Hardware, Software, and Services behind the New Manufacturing Revolution*. O'Reilly & Associates Inc., 2016.
This is a guide for the 3D printing newbie. It begins with the history of 3D printing, then reviews various printers and software, and features some tutorials. Not necessarily for educators but valuable general information anyway.

Otfinoski, Steven. *3D Printing: Science, Technology, Engineering*. Children's Press, 2017.
This provides a review of today's innovations in the field of 3D printing. The possibility of 3D printing not too long ago seemed like a fantasy. Today, companies are working on making printers more affordable and faster, as well as seamlessly connected to your desktop or tablet. There is also a review of the future of using 3D printing in serious manufacturing as well as far-fetched and unheard-of uses for 3D printing.

Chapter 5

ENGINEERING FOR KIDS?

THE CASE FOR ENGINEERING FOR KIDS

- Consistently, kids think engineers drive trains and build bridges.
- Kids need to know that engineers DESIGN buildings, bridges, vehicles, and the like.
- Engineering can provide opportunities for kids to work together to solve problems at an early age.
- Applying engineering skills to real-world scenarios better prepares students for careers.
- Problem-solving skills are a very important part of engineering.
- Students need to learn how to face a challenge by trying, failing, working on, rethinking, and trying again.
- Failure is an important part of the design process, and there is no single right answer.
- Engineering activities require students to work in teams and collaborate.

Engineering in the lower grades is probably the most difficult topic to introduce because it seems better suited to older students. There are not many nonfiction or reference books that provide information and activities suitable for the PK-5 crowd in the field of engineering.

One of the biggest difficulties with the subject of engineering appears to be the fact that most kids do not know exactly what an engineer does. The possibilities for a career in engineering are endless, but many think of an engineer as someone who drives a train or builds bridges. They do not realize the importance of the design element used by engineers. The job can be very diverse, and a comprehensive list of all the possibilities in the field of engineering can be found here: http://www.egfi-k12.org/engineer-your-path/

Engineers are problem-solvers, and they have a part in one way or another in almost everything in our daily lives. And they are instrumental in making changes and improvements for the future. The Museum of Science in Boston has a program and website that has wonderful resources for helping kids understand the importance of engineering. Here is a link to a trailer for an IMAX film called *Dream Big*, narrated by Jeff Bridges about the wonders of engineering: https://www.mos.org/imax/dream-big. And Engineering Is Elementary (https://eie.org/) offers curriculum for multiple grade levels for sale, but the website has free resources as well.

ANNOTATED JOURNAL ARTICLES

Gorman, Dorothy Powers. "Engineering in Elementary Schools: Engaging the Next Generation of Problem Solvers Today." *South East Education Network* 16:3 (2014): 36–37.
   Research shows that most young kids do not have a clear understanding of engineering. The author has learned that early on students decide whether or not they like or are good at science and math. So, teachers have a job to do with regard to supplying information and encouraging interest. The concepts of problem-solving, critical thinking, and innovation can easily be shared with young students. We know that they are capable of understanding engineering design concepts at an early age. It is especially important for minorities and girls to be enticed into the world of problem-solving in order to increase the number of students selecting engineering as a career. It is not only an equity issue; it is also an economic issue as engineers are in short supply.

Cunningham, Christine M. and Melissa Higgins. "Engineering for Everyone." *Educational Leadership* 72:4 (2014–2015): 42–47.
   These authors, who both have important positions with the Engineering Is Elementary program at the Museum of Science in Boston (https://eie.org/), state that the best way to introduce engineering to young students is through tinkering and makerspaces. The makerspaces for such activities for this age group can be a center/workstation format with very simplistic materials. The most important part of this is providing a real-world problem that works for kids in this age group. They also think that it is more meaningful to present a problem in story format and make it one with real-world connections. They also want engineers to be viewed as people who help others and that sometimes it takes many tries and failures before a problem is solved. Engineers work together in teams to design and build things that make our lives easier. It is helpful if the teacher structures the hands-on activities with multiple possible solutions. And it is not necessary to have costly tools and kits to teach engineering. The design principles can be taught using readily available and inexpensive or even recycled materials.

"Elementary Engineering: From Simple Machines to Life Skills." *Edutopia*. January 26, 2016. https://www.edutopia.org/practice/elementary-engineering-simple-machines-life-skills
   The Charles R. Drew Charter School in Atlanta, Georgia, has developed an engineering curriculum using backwards design beginning with the youngest students. The Next Generation Science Standards have included some engineering design skills even in the elementary grades. Students have to know the value of problem-solving using teamwork. They have to be aware that they may try a number of possible solutions before they get the one that works and that perseverance is important and valued.

Students begin by making a plan, which includes the goals. Next, they learn all they can about the problem or challenge—using the library, the Internet, and experts. Then they try different approaches based on their research and eventually create a prototype. Next, they evaluate and make adjustments until they get the desired results. They then share their results with peers and the community. They use a rubric based on following the design process as well as a student evaluation form regarding the process and the results.

Borovoy, Amy Erin. "5-Minute Film Festival: Learning with Rube Goldberg Machines." *Edutopia.* June 24, 2016. https://www.edutopia.org/film-fest-rube-goldberg-learning-ideas
   Rube Goldberg was first an engineer and then a cartoonist. Borovoy has created a video playlist of a variety of Rube Goldberg inventions to inspire young people to try to create a crazy invention to get a relatively simple problem solved. Engineering design, problem-solving, perseverance, an understanding of simple machines, and a sense of humor all figure into a Rube Goldberg invention!

Abrams, Michael. "Bot for Tots." American Society of Mechanical Engineers. February 2014. https://www.asme.org/career-education/articles/k-12-grade/bots-for-tots
   Abrams reports on the research and development of a pair of small robots designed for some youngsters. The object is to teach the kids how to program the robots realizing that they cannot type or therefore use text or code to program. In a world where today just about everything is programmed, kids need to start young in understanding the processes. These robots were designed using play and music. The interfaces change for older kids.

Carmody, Erin. "Running an Engineering Design Challenge—5 Tips to Get Anyone Started." STEM Village. October 12, 2016. http://www.stemvillage.com/running-an-engineering-design-challenge-5-tips-to-to-get-anyone-started/
   Carmody is the content manager for STEM Village, an online program with all sorts of resources for learning and teaching about STEM and engineering. Here she talks about a design challenge and shares steps for organizing the process. The plan can be for a whole school, a grade level, or a classroom; and the procedures are understandable and easy to implement. The most important aspect is to keep the challenge broad: a variety of stakeholders, a variety of possible outcomes, and to make it appealing to the participants. Also, allow time and place for feedback. The makerspace does not have to be filled with sophisticated tools and equipment—she even lists easy to get and inexpensive supplies.

ENGINEERING ACTIVITY BOOKS

Andrews, Beth L. *Hands-on Engineering: Real-World Projects for the Classroom.* Prufrock Press Inc., 2012. ISBN: 978-1-593-639228. Gr 4-7

Beaty, Andrea. *Ada Twist's Big Project Book for Stellar Scientists.* Harry N. Abrams, 2017. ISBN: 978-1-419-730245. Gr K-2

Beaty, Andrea. *Iggy Peck's Big Project Book for Amazing Architects.* Harry N. Abrams, 2017. ISBN: 978-1-683-351306. Gr K-2

Beaty, Andrea. *Rosie Revere's Big Project Book for Bold Engineers.* Harry N. Abrams, 2017. ISBN: 978-1-613-125304. Gr K-2

Ceceri, Kathy. *Making Simple Robots: Exploring Cutting-Edge Robotics with Everyday Stuff.* Maker Media, 2015. ISBN: 978-1-457-183638. Gr 6-12

Ceceri, Kathy and Samuel Carbaugh. *Robotics: Discover the Science and Technology of the Future with 20 Projects.* Nomad, 2012. ISBN: 978-1-936-749751. Gr 3-7

DK *Find Out! Engineering.* DK, 2017. ISBN: 978-1-465-462343. Gr 1-4

Jacoby, Jenny and Vicky Barker. *STEM Starters for Kids Engineering Activity Book: Packed with Activities and Engineering Facts.* Racehorse for Young Readers, 2017. ISBN: 978-1-631-581946. Gr 2-3

McCue, Camille. *Getting Started with Engineering: Think Like an Engineer.* John Wiley, 2016. ISBN: 978-1-119-291220. Gr 2-5

Reyes, Sandi. *Engineer through the Year: 20 Turnkey STEM Projects to Intrigue, Inspire & Challenge—Grades K-2.* SDE Crystal Springs Books, 2012. ASIN: 1935502379. Gr K-2

Reyes, Sandi. *Engineer through the Year: 20 Turnkey STEM Projects to Intrigue, Inspire & Challenge—Grades 3–5.* SDE Crystal Springs Books, 2012. ASIN: B00QM20SFC. Gr 3-5

Chapter 6

SCIENCE

To raise new questions, new possibilities, to regard old problems from a new angle, requires creative imagination and marks real advance in science.

—Albert Einstein
http://www.brainyquote.com

Science *includes the nature of concepts, processes, inquiry, and the traditional subjects of physics, biology, chemistry, space and geosciences, and biochemistry. Science deals with the understanding of the natural world and is the underpinning of technology. There is a misconception that technology is applied science but instead it is important for students to develop a greater understanding and appreciation for some of the fundamental concepts and processes of technology and engineering. Science is concerned with what is in the natural world and some of the processes include inquiry, discovery, exploration, and using the scientific method.*

—Dugger, William E. "STEM: Some Basic Definitions"
(Senior Fellow, International Technology and
Engineering Educators Association).
http://www.iteea.org

GENERAL QUESTIONS

What did you learn from this book?

What qualifications did the author have to write this book?

Where can you go to find out more information about this topic?

What was the purpose of this book?

What does the author want the reader to believe about this topic?

Was there anything in this book that you did not understand?

FEATURED AUTHORS AND ANNOTATIONS

Name: Steve Jenkins

Place of Birth: Hickory, North Carolina

About: His father was in the military, and the family moved often. As a child, he kept a small menagerie of lizards, turtles, spiders, and other animals. He collected rocks and fossils and blew up things in his small chemistry lab. He went to school to study graphic design and eventually opened his own graphic design business.

Website: http://www.stevejenkinsbooks.com

Annotated Title:

Jenkins, Steve. *Eye to Eye: How Animals See the World*. Houghton Mifflin Harcourt, 2014. ISBN: 978-0-54795-907-8. Gr 2-6

> "Being able to see helps animals communicate, find food, avoid predators, or locate a mate." There are four kinds of eyes: eyespot, pinhole eyes, compound eyes, and camera eye. Jenkins begins with the simplest eye of the sea slug and ends with the sharpest eyesight of the Eurasian Buzzard, which can hone in on a rabbit two miles away. This book includes the evolution of the eye, animal facts, and a glossary.

Name: Sandra Markle

Place of Birth: Fostoria, Ohio

About: She has published more than 200 nonfiction books for children, has won numerous awards for this work, including the 2012 Prize for Excellence in Science Books by the American Association for the Advancement of Science (AAAS), *Boston Globe-Horn Book Honor Book*, NSTA Outstanding Science Tradebook, Green Earth Book Awards, Orbis Pictus Recommended Book, and Charlotte Zolotow Award, among others.

Website: http://sandra-markle.blogspot.com/

Annotated Title:

Markle, Sandra. *The Case of the Vanishing Honeybees: A Scientific Mystery*. Lerner Publishing Group, 2013. ISBN: 978-1-4677-0592-9. Gr 4-8

> Honeybees and plants depend on each other. "As they gather nectar from flowers to make sweet honey, these bees also play an important role in pollination, helping some plants produce fruit." Since 2007, beekeepers in many countries were losing 30% to 50% of their hives. Many theories are explored while scientists are trying to find a way to stop the decline.

Name: Seymour Simon

Place of Birth: New York City, New York

About: He has written nearly 300 books. He was once a science teacher in a New York City middle school. When his first book was published, he felt just like you all do when you have done something that you are really proud of and cannot wait to show it to your parents. He wrote a dedication in it to his mother and took it straight over to show to her. The *New York Times* called Seymour Simon "the dean of children's science."

Website: http://www.seymoursimon.com
Annotated Title:
Simon, Seymour. *Coral Reefs*. HarperCollins, 2013. ISBN: 978-0-191495-9. Gr 2-5
This book describes the different types of coral reefs and their inhabitants with great photos and simple language. Simon tells about the environmental changes, such as global warming and toxic waste, that are harming the reefs. There is a helpful glossary included.

ACTIVITIES THAT CONNECT SCIENCE

Honeybees

Science	Technology	Engineering
Markle, Sandra. *The Case of the Vanishing Honeybees: A Scientific Mystery*. Lerner Publishing Group, 2014. Gr 4-8	The Waggle and the Web: Honeybee Technology from *How Stuff Works*. http://animals.howstuffworks.com/insects/honeybee-web-server1.htm	How to Make a Honey Bee Box. http://www.wikihow.com/Make-a-Honey-Bee-Box

Arts	Math
Draw honeycomb (tangle pattern)—https://www.youtube.com/watch?v=5jEUPquHw_4	Hexagons—http://www.mathopenref.com/hexagon.html

INVENTORS AND INVENTIONS

Jones, Charlotte. *Mistakes That Worked*. Delacorte Books for Young Readers, 1994. ISBN: 978- 0385320436. Gr 3-7

Oxlade, Chris. *The Camera*. Heinemann Library, 2010. ISBN: 978-0-43111-841-3. Gr 3-5

Spengler, Kremena T. *An Illustrated Timeline of Inventions and Inventors*. Picture Window Books, 2011. ISBN: 978-1404870178. Gr 2-4

Wulffson, Don. *The Kid Who Invented the Popsicle: And Other Surprising Stories about Inventions*. Puffin, 1999. ISBN: 978-0-14130-204-1. Gr 3-7

Wulffson, Don. *Toys! Amazing Stories behind Some Great Inventions*. Henry Holt and Co., 2000. ISBN: 978-0-805-06196-3. Gr 3-7

BIBLIOGRAPHY

Life Science: Definition

A branch of science (such as biology, medicine, and sometimes anthropology or sociology) that deals with living organisms and life processes—usually used in plural.

—Merriam-Webster

Picture Books: Life Science

Applegate, Katherine. *Ivan: The Remarkable True Story of the Shopping Mall Gorilla*. Clarion Books, 2014. ISBN: 978-0-54425-230-1. Gr PreS-3

Aston, Dianna Hutts. *An Egg Is Quiet*. Chronicle Books, 2014. ISBN: 978-1-452-131481. Gr K-3

Austin, Mike. *Junkyard*. Beach Lane Books, 2014. ISBN: 978-1-442-459618. Gr PreS-3

Beaty, Andrea. *Ada Twist, Scientist*. Abrams Books for Young Readers, 2016. ISBN: 978-1419721373. Gr K-2

Butler, Chris. *See What a Seal Can Do*. Candlewick, 2013. ISBN: 978-0-763-66574-6. Gr K-4

Cate, Annette LeBlanc. *Look Up! Bird-Watching in Your Own Backyard*. Candlewick, 2013. ISBN: 978-0-76364-561-8. Gr 2-4

Cherry, Lynne. *The Great Kapok Tree: A Tales of the Amazon Rain Forest*. HMH Books for Young Readers, 1990. ISBN: 978-0-15200-520-7. Gr PreS-3 Classic

Chin, Jason. *Island: A Story of the Galapagos*. Roaring Brook, 2012. ISBN: 978-1-59643-716-6. Gr K-3

Cole, Joanna. *The Magic School Bus and the Science Fair Expedition*. Scholastic, 2006. ISBN: 0-590-10824-7. Gr K-3

Coleman, Janet Wyman. *Eight Dolphins of Katrina: A True Tale of Survival*. HMH Books for Young Readers, 2013. ISBN: 978-0-547-71923-8. Gr 1-4

Davies, Nicola. *Outside Your Window: A First Book of Nature*. Candlewick Press, 2012. ISBN: 978-0-76365-549-5. Gr PreS-2

Elliott, David. *On the Wing*. Candlewick, 2014. ISBN: 978-076-365-3248. Gr K-3

George, Jean Craighead. *The Eagles Are Back*. Dial, 2013. ISBN: 978-0-803-73771-6. Gr 1-3

George, Jean Craighead. *The Wolves Are Back*. Dutton Books for Young Readers, 2008. ISBN: 978-0-525-479475. Gr PreS-3

Gravel, Elise. *The Fly (Disgusting Critters Series)*. Tundra Books, 2014. ISBN: 978-1-77049-636-1. Gr K-3

Gray, Leon. *Giant Pacific Octopus: The World's Largest Octopus*. Bearport Publishing, 2013. ISBN: 978-1-617-72730-6. Gr 1-3

Guiberson, Brenda Z. *Feathered Dinosaurs*. Henry Holt and Co., 2016. ISBN: 978-0-805-098280. Gr K-3

Guiberson, Brenda Z. *Frog Song*. Henry Holt and Co., 2013. ISBN: 978-0-805-09254-7. Gr PreS-3

Guiberson, Brenda Z. *Life in the Boreal Forest*. Henry Holt and Co., 2009. ISBN: 978-0-805-07718-6. Gr K-4

Hatkoff, Juliana. *Knut: How One Little Polar Bear Captivated the World*. Scholastic Press, 2007. ISBN: 978-0-54504-716-6. Gr K-3

Heinz, Brian. *Mocha Dick: The Legend and the Fury*. Creative Editions, 2014. ISBN: 987-1-568-462-424. Gr 3 and Up

Himmelman, John. *A Monarch Butterfly's Life*. Children's Press, 2000. ISBN: 978-0-516-265377. Gr K-3

Hollard, Mary. *Animal Mouths*. Arbordale Publishing, 2015. ISBN: 978-1-628-555615. Gr 1-4

HoPreSins, H. Joseph. *The Tree Lady: The True Story of How One Tree-Loving Woman Changed a City Forever*. Beach Lane Books, 2013. ISBN: 978-1-44241-402-0. Gr K-5

Huber, Raymond. *Flight of the Honey Bee*. Candlewick, 2013. ISBN: 978-0-763-66760-3. Gr PreS-2

Jenkins, Steve. *Almost Gone: The World's Rarest Animals*. HarperCollins, 2006. ISBN: 978-0-060-536008. Gr K-4

Jenkins, Steve. *Animals by the Numbers*. HMH Books for Young Readers, 2016. ISBN: 978-0-544-630-925. Gr 1-4

Jenkins, Steve. *Animals Upside Down*. HMH Books for Young Readers, 2013. ISBN: 978-05-547-34127-9. Gr PreS-3

Jenkins, Steve. *Born in the Wild: Baby Mammals and Their Parents*. Roaring Brook Press, 2014. ISBN: 978-1-59643-925-2. Gr K-3

Jenkins, Steve. *Creature Features: Twenty-Five Animals Explain Why They Look the Way They Do*. HMH Books for Young Readers, 2014. ISBN: 978-0-54423-351-5. Gr PreS-3

Jenkins, Steve. *Egg: Nature's Perfect Package*. HMH Books for Young Readers, 2015. ISBN: 978-0-547-959092. Gr PreS-3

Jenkins, Steve. *Flying Frogs and Walking Fish*. HMH Books for Young Readers, 2016. ISBN: 978-0-544-630901. Gr K-3

Jenkins, Steve. *How Many Ways Can You Catch a Fly?* HMH Books for Young Readers, 2008. ISBN: 978-0-61896-634-9. Gr PreS-3

Jenkins, Steve. *What Do You Do with a Tail Like This?* HMH Books for Young Readers, 2009. ISBN: 978-0-439-703840. Gr PreS-3

Judge, Lita. *Bird Talk: What Birds Are Saying and Why*. Roaring Brook Press, 2012. Gr 2-4

Lawrence, Ellen. *A Dragonfly's Life*. Bearport Publishing, 2012. ISBN: 978-1-617-72594-4. Gr 1-4

Lawrence, Ellen. *Meat-Eating Plants: Toothless Wonders*. Bearport Publishing, 2012. ISBN: 978-1-617-72589-0. Gr PreS-3

London, Jonathan. *Otters Love to Play*. Candlewick, 2016. ISBN: 978-0-763-669133. Gr K-3

Lurie, Susan. *Swim, Duck, Swim!* Feiwel & Friends, 2014. ISBN: 978-1-25004-642-0. Gr PreS-2

Markle, Sandra. *Growing Up Wild: Penguins*. Atheneum, 2002. ISBN: 0-689-81887-4. Gr K-3

Markle, Sandra. *The Long, Long Journey: The Godwit's Amazing Migration*. Millbrook, 2013. ISBN: 978-0-76135-623-3. Gr 1-3

Markle, Sandra. *Outside and Inside Sharks*. Aladdin, 1999. ISBN: 978-0-6898-2683-2. Gr K-3

Markle, Sandra. *Snow School*. Charlesbridge, 2013. ISBN: 978-1-580-89410-4. Gr PreS-3

Markle, Sandra. *What If You Had Animal Feet?* Scholastic Paperback Nonfiction, 2015. ISBN: 978-0-54573-312-0. Gr K-3

Markle, Sandra. *What If You Had Animal Teeth?* Scholastic Paperbacks, 2013. ISBN: 978-0-54548-438-1. Gr PreS-3

Marsh, Laura. *National Geographic Readers: Great Migrations Butterflies*. National Geographic Children's Books, 2010. ISBN: 978-1-426-307409. Gr 2-4

Momatiuk, Yva. *Face to Face with Penguins*. National Geographic Children's Book, 2009. ISBN: 978-1-426-30561-0. Gr 2-5

Momatiuk, Yva. *Face to Face with Wild Horses*. National Geographic Children's Books, 2009. ISBN: 978-0-426-30466-8. Gr 2-5

Morgan, Emily. *Next Time You See a Firefly*. National Science Teachers Association, 2013. ISBN: 978-1-936-95918-1. Gr 1-3

National Geographic Learning. *Where Do Frogs Come From?* National Geographic School Pub, 2012. ISBN: 978-0-152-048440. Gr K-2

Offill, Jenny. *Eleven Experiments That Failed*. Random House, 2011. ISBN: 978-0-375-84762-2. Gr K-2

Person, Stephen. *Cougar*. Bearport Publishing, 2012. ISBN: 978-1-617-72569-2. Gr 3-5

Person, Stephen. *Roseate Spoonbill: Pretty in Pink*. Bearport Publishing, 2012. ISBN: 978-1-617-72570-8. Gr 3-5

Pfeffer, Wendy. *Life in a Coral Reef*. HarperCollins, 2009. ISBN: 978-0-06-029553-0. Gr K-3

Porter, Esther. *Peeking under the City*. Capstone Press, 2016. ISBN: 978-1-479-586653. Gr K-2

Rotner, Shelley. *Yummy! Good Food Makes Me Strong*. Holiday House, 2013. ISBN: 978-0-823-42426-9. Gr PreS-1

Rotner, Shelley. *Grow! Raise! Catch!* Holiday House, 2016. ISBN: 978-0-823-438846. Gr PreS-1

Roy, Katherine. *Neighborhood Sharks: Hunting with the Great Whites of California's Farallon Islands*. David Macaulay Studio, 2014. ISBN: 978-1-59643-874-3. Gr 2-6

Sayre, April Pulley. *Eat Like a Bear*. Henry Holt and Co., 2013. ISBN: 978-0-805-09039-0. Gr PreS-3

Sayre, April Pulley. *Here Come the Humpbacks*. Charlesbridge, 2013. ISBN: 978-1-580-89406-7. Gr PreS-3

Schaefer, Lola M. *Lifetime: The Amazing Numbers in Animal Lives*. Chronicle Books, 2013. Gr K-4

Scieszka, Jon. *Science Verse*. Viking, 2004. ISBN: 0-670-91057-0. Gr K-3 Classic

Simon, Seymour. *Insects*. HarperCollins Children's Books, 2016. ISBN: 978-0-062-289155. Gr 1-5

Stevenson, Emma. *Hide-And-Seek Science: Animal Camouflage*. Holiday House, 2013. ISBN: 978-0-823-42293-7. Gr 2-5

Stockdale, Susan. *Stripes of All Types*. Peachtree Publishers, 2013. ISBN: 978-1-561-45695-6. Gr PreS-1

Stone, Tanya Lee. *Who Says Women Can't Be Doctors? The Story of Elizabeth Blackwell*. Henry Holt and Co., 2013. ISBN: 978-0-805-09048-2. Gr K-3

Thermes, Jennifer. *Charles Darwin's Around-the-World Adventure*. Abrams Books for Young Readers, 2016. ISBN: 978-1-419-721205. Gr K-3

Winter, Jeanette. *The Watcher: Jane Goodall's Life with the Chimps*. Schwartz & Wade, 2011. ISBN: 978-0-37586-774-3. Gr PreS-3

Informational: Life Science

Albee, Sarah. *Bugged: How Insects Changed History*. Walker, 2014. ISBN: 978-0-80273-422-8. Gr 4-9

Aldefer, Jonathan. *National Geographic Kids Bird Guide of North America: The Best Birding Book for Kids from National Geographic Bird Experts*. National Geographic Children's Books, 2013. ISBN: 978-1-426-31095-9. Gr 2-4

Arndt, Ingo. *Animal Architecture*. Harry N. Abrams, 2014. ISBN: 978-1-419-711657. Reference

Arnold, Caroline. *Too Hot? Too Cold? Keeping Body Temperature Just Right*. Charlesbridge, 2013. ISBN: 978-1-58089-277-3. Gr 1-4

Bardoe, Cheryl. *Mammoths and Mastodons: Titans of the Ice Age*. Abrams, 2010. ISBN: 978-0-8109-8413-4. Gr 4-6

Becker, Tom. *A Zombie's Guide to the Human Body*. Scholastic Reference, 2010. ISBN: 978-0-5452-4979-9. Gr 3-7

Bennett, Jeffrey. *A Global Warming Primer*. Big Kid Science, 2016. ISBN: 978-1-937-548780. YA

Bennett, Jeffrey. *I, Humanity*. Big Kid Science, 2016. ISBN: 978-1-937-548520. Gr 2-5

Brennan, Linda Crotta. *When Rivers Burned: The Earth Day Story*. Apprentice Shop Books, 2013. ISBN: 978-0-984-25499-6. Gr 7 and Up

Brusatte, Steve. *Day of the Dinosaurs*. Wide Eyed Editions, 2016. ISBN: 978-1-847-808455. Gr 2-5

Burns, Loree Griffin. *Citizen Scientists: Be a Part of Scientific Discovery from Your Own Backyard*. Henry Holt, 2012. ISBN: 978-0-8050-9062-8. Gr 3-5

Burns, Loree Griffin. *The Hive Detectives: Chronicle of a Honey Bee Catastrophe*. HMH Books for Young Readers, 2010. ISBN: 978-0-547-152318. Gr 5-8

Cerullo, Mary M. *Giant Squid*. Capstone Press, 2014. ISBN: 978-1-42967-541-3. Gr 4-7

Cobb, Vicki. *Your Body Battles a Stomach Ache*. Millbrook Press, 2009. ISBN: 978-0-822-571667. Gr 2-5

Collard, Sneed B., III. *Hopping ahead of Climate Change*. Bucking Horse Books, 2016. ISBN: 978-0-984-446-087. Gr 2-5

Cusick, Dawn. *Get the Scoop on Animal Poop: From Lions to Tapeworms: 251 Cool Facts about Scat, Frass, Dung, and More.* Imagine Publishing, 2012. ISBN: 978-1-62354-104-2. Gr 2-5

Donovan, Sandra. *Does It Really Take Seven Years to Digest Swallowed Gum? And Other Questions You've Always Wanted to Ask.* Lerner, 2010. ISBN: 978-0-8225-9085-9. Gr 4-6

Evans, Michael and David Wichman. *The Adventures of Medical Man: Kids Illnesses and Injuries Explained.* Annick, 2010. ISBN: 978-1-55451-263-8. Gr 5-8 Graphic Novel

Fleischman, Paul. *Eyes Wide Open: Going behind the Environmental Headlines.* Candlewick, 2014. ISBN: 978-0-76367-545-5. YA

Fleming, Candace. *Giant Squid.* Roaring Brook Press, 2016. ISBN: 978-1-596-435995. Gr 2-4

Goldsmith, Connie. *Traumatic Brain Injury.* Twenty-First Century, 2014. ISBN: 978-1-46771-348-1. Gr 7 and Up

Goodman, Susan E. *Gee Whiz! It's All about Pee.* Viking, 2006. ISBN: 978-0-670-06064-1. Gr 3-5

Goodman, Susan E. *The Truth about Poop.* Puffin; Reprint Edition, 2007. ISBN: 978-0-1424-0930-5. Gr 3-5

Griffiths, Andy. *What Body Part Is That? A Wacky Guide to the Funniest, Weirdest, and Most Disgustingest Parts of Your Body.* Feiwel and Friends, 2011. ISBN: 978-0-312-36790-9. Gr 3-7

Haeckel, Ernst. *Art Forms in Nature: The Prints of Ernst Haeckel.* Prestel, 2008. ISBN: 978-3-791-319902. Reference

Hearst, Michael. *Unusual Creatures: A Mostly Accurate Account of Some of Earth's Strangest Animals.* Chronicle, 2012. ISBN: 978-1-45210-467-6. Gr 2-6

Hendrick, Gail. *Something Stinks.* Tumblehome Learning, Inc., 2013. ISBN: 978-0-985-00089-9. Gr 4-7 Outstanding Science Trade Book for Students K-12 for 2014 (Fiction)

Hoose, Phillip. *Moonbird: A Year on the Wind with the Great Survivor B95.* Farrar, Straus and Giroux, 2012. ISBN: 978-0-37340-468-3. Gr 4-8

Howowitz, Alexandra. *Inside of a Dog.* Simon & Schuster Books for Young Readers, 2016. ISBN: 978-1-481-450935. Gr 3 and Up

Ignotofsky, Rachel. *Women in Science: 50 Fearless Pioneers Who Changed the World.* Ten Speed Press, 2016. ISBN: 978-1-607-749769. YA

Jarrow, Gail. *Red Madness: How a Medical Mystery Changed What We Eat.* Calkins Creek, 2014. ISBN: 978-1-59078-732-8. Gr 5 and Up

Jenkins, Steve. *The Animal Book: A Collection of the Fastest, Fiercest, Toughest, Cleverest, Shyest–and–Most Surprising–Animals on Earth*. Houghton Harcourt, 2013. ISBN: 978-54755-799-1. Gr 2-6

Jenkins, Steve. *Bones: Skeletons and How They Work*. Scholastic, 2010. ISBN: 978-0-54504-651-0. Gr 3-6

Jenkins, Steve. *Eye to Eye: How Animals See the World*. Houghton Harcourt, 2014. ISBN: 978-054-7959-078. Gr 2-6

Jenkins, Steve. *How to Swallow a Pig: Step-by-Step Advice from the Animal Kingdom*. HMH Books for Young Readers, 2015. ISBN: 978-0-544-313651. Gr 2-5

Johnson, Rebecca L. *Journey into the Deep: Discovering New Ocean Creatures*. Millbrook, 2010. ISBN: 978-0-7613-4148-2. Gr 4-6

Johnson, Rebecca L. *Zombie Makers: True Stories of Nature's Undead*. Millbrook Press, 2012. ISBN: 978-0-7613-8633-9. Gr 5-8

Lin, Grace. *Our Food*. Charlesbridge, 2016. ISBN: 978-1-580-895903. Gr 3-4

Lourie, Peter. *Whaling Season: A Year in the Life of an Arctic Whale Scientist*. Houghton Mifflin, 2009. ISBN: 978-0-618-77709-9. Gr 4-8

Macaulay, David. *Eye: How It Works*. Roaring Brook, 2013. ISBN: 978-1-59643-782-1. Gr PreS-1

Macaulay, David. *The Way We Work: Getting to Know the Amazing Human Body*. HMH Books for Young Readers, 2008. ISBN: 978-061-8233-786. Gr 5 and Up

Maggs, Sam. *Wonder Women*. Quirk Books, 2016. ISBN: 978-1-594-749254. YA

Marsh, Laura F. *Caterpillar to Butterfly*. National Geographic, 2012. ISBN: 978-1-4263-0920-6. Gr PreS-2

Martin, Jacqueline Briggs. *Farmer Will Allen and the Growing Table*. Readers to Eaters, 2013. ISBN: 978-0-98366-153-5. Gr 3-6

Mills, Andrea. *True or False*. DK, 2014. ISBN: 978-1-46542-467-9. Gr 3-8

Monroe, Randall. *What If: Serious Scientific Answers to Absurd Hypothetical Questions*. HMH, 2014. ISBN: 978-0-54427-299-6. Gr 8 and Up

Montgomery, Sy. *The Great White Shark Scientist*. HMH Books for Young Readers, 2016. ISBN: 978-0-544-352988. Gr 5-8

Montgomery, Sy. *Kakapo Rescue*. HMH Books for Young Readers, 2016. ISBN: 978-0-544-668294. Gr 5-7

Montgomery, Sy. *Quest for the Tree Kangaroo: An Expedition to the Cloud Forest of New Guinea*. HMH Books for Young Readers, 2009. ISBN: 978-0-547-24892-2. Gr 5-9

Montgomery, Sy. *The Tarantul Scientist*. HMH Books for Young Readers, 2007. ISBN: 978-0-618-915744. Gr 5-7

Murawski, Darlyne. *Ultimate Bugopedia*. National Geographic Kids, 2013. ISBN: 978-1-426-31376-9. Gr 2-6

Murphy, Glenn. *How Loud Can You Burp? More Extremely Important Questions (and Answers!)* Flash Point Paper, 2009. ISBN: 978-1-59643-506-3. Gr 4-7

Murphy, Glenn. *Stuff That Scares Your Pants Off! The Science Scoop on More than 30 Terrifying Phenomena*. Roaring Book Press, 2011. ISBN: 978-1-5964-3633-6. Gr 6-9

Newman, Patricia. *Plastic Ahoy! Investigating the Great Plastic Garbage Patch*. Millbrook Press Trade, 2014. ISBN: 978-1-467-712835. Gr 5-8

O'Connell, Caitlin. *A Baby Elephant in the Wild*. Houghton Harcourt, 2014. ISBN: 978-0-54414-944-1. Gr PreS-3.

O'Meara, Stephen Jones. *Are You Afraid Yet? The Science behind Scary Stuff*. Kids Can Press, 2009. ISBN: 978-1-5545-3294-0. Gr 6-9

Ottavianni, Jim. *Primates: The Fearless Science of Jane Goodall, Dian Fossey, and Birute Galdikas*. First Second Books, 2013. ISBN: 978-1-5964-3865-1. Gr 7 and Up

Parker, Nancy Winslow. *Organs: How They Work, Fall Apart, and Can Be Replaced (Gasp!)*. Greenwillow, 2009. ISBN: 978-0-68815-105-8. Gr 2-5

Parker, Steve. *Super Human Encyclopedia*. Dorling Kindersley, 2014. ISBN: 978-1-46542-445-7. Gr 6 and Up

Pinnington, Andrea. *Scholastic Discover More: My Body*. Scholastic, 2012. ISBN: 978-0-54534-514-9. Gr PreS-3

Pringle, Laurence. *Owls*. Boyds Mills Press, 2016. ISBN: 978-1-620-916513. Gr 3-6

Rhatigan, Joe. *Ouch: The Weird and Wild Ways Your Body Deals with Agonizing Aches, Ferocious Fevers, Lousy Lumps, Crummy Colds, Bothersome Bites, Breaks, Bruises and Burns and Makes Them Feel Better*. Imagine, 2013. ISBN: 978-1-62354-005-0. Gr 3-7

Roth, Susan L. and Cindy Trumbore. *Parrots over Puerto Rico*. Lee & Low, 2013. ISBN: 978-1-62014-004-8. Gr 3-6

Rotner, Shelley. *Body Bones*. Holiday House, 2014. ISBN: 978-0-82343-162-8-5. Gr 1-3

Sidman, Joyce. *Winter Bees and Other Poems of the Cold*. Houghton Mifflin Harcourt, 2014. ISBN: 978-0-54790-650-8. Gr K-4

Sis, Peter. *The Tree of Life: A Book Depicting the Life of Charles Darwin, Naturalist, Geologist and Thinker.* Farrar, Straus and Giroux, 2003. ISBN: 978-0-374-456283. Gr 5-8

Sloan, Christopher. *Tracking Tyrannosaurs: Meet T. Rex's Fascinating Family, from Tiny Terrors to Feathered Giants.* National Geographic Children's Books, 2013. ISBN: 978-1-426-31374-5. Gr 3-7

Soloway, Andrew. *Sports Science.* Heinemann-Raintree, 2009. ISBN: 978-1-4329-2480-5. Gr 4-9

Squarzoni, Philippe. *Climate Changed: A Personal Journey through the Science.* Harry N. Abrams, 2014. ISBN: 978-1-419-712555. Gr 7 and Up

Stewart, Melissa. *A Place for Turtles.* Peachtree Publishers, 2013. ISBN: 978-1-561-45693-2. Gr 1-4

Swanson, Jennifer. *Super Gear.* Charlesbridge, 2016. ISBN: 978-1-580-897204. Gr 6-9

Swinburne, Stephen R. *Sea Turtle Scientist.* Houghton Harcourt, 2014. ISBN: 978-054-7367-552. Gr 2-6

Szpirglas, Jeff. *You Just Can't Help It: Your Guide to the Wild and Wacky World of Human Behavior.* Owlkids Books, 2011. ISBN: 978-1-92681-808-5. Gr 3-7

Thimmesh, Catherine. *Lucy Long Ago: Uncovering the Mystery of Where We Came From.* Houghton Mifflin Books for Children, 2010. ISBN: 978-0-54705-199-4. Gr 5 and Up

Thimmesh, Catherine. *Scaly Spotted Feathered Frilled: How Do We Know What Dinosaurs Really Looked Like?* Houghton Harcourt, 2013. ISBN: 978-0-54799-134-4. Gr 5-7

Thornhill, Jan. *The Tragic Tale of the Great Auk.* Groundwood Books, 2016. ISBN: 978-1-554-988665. Gr 2-6

Time Inc. Books. *Strange, Unusual, Gross and Cool Animals.* Animal Planet, 2016. ISBN: 978-1-618-931665. Gr 3-5

Turner, Pamela S. *The Dolphins at Shark Bay.* Houghton Mifflin, 2013. ISBN: 978-0-54771-638-1. Gr 5-9

Wadsworth, Ginger. *Poop Detectives.* Charlesbridge, 2016. ISBN: 978-1-580-896503. Gr 3-7

Walker, Sally M. *Written in Bone: Buried Lives of Jamestown and Colonial Maryland.* Carolrhodda Books, 2009. ISBN: 978-0-82257135-3. Gr 6-12

Wicks, Maris. *Human Body Theater.* First Second, 2015. ISBN: 978-1-626-722773. Gr 4-8

Wood, Amanda. *Natural World.* Wide Eye Editions, 2016. ISBN: 978-1-847-807823. Gr 3 and Up

Yoder, Eric. *One Minute Mysteries: 65 More Short Mysteries You Solve with Science.* Platypus Media, 2012. ISBN: 978-1-938-49200-9-6. Gr 3-7

Zemlicka, Shannon. *Florence Nightingale.* Lerner Publishing Group, 2003. ISBN: 978-0-87614-102-1. Gr 2 and Up

Earth Science: Definition

Any of the sciences (such as geology, meteorology, or oceanography) that deal with the earth or with one of its parts—compare geoscience.

—Merriam-Webster

Picture Books: Earth Science

Adler, David A. *Things That Float and Things That Don't.* Holiday House, 2013. ISBN: 978-0-823-42862-5. Gr K-2

Bahr, Mary. *My Brother Loved Snowflakes: The Story of Wilson A. Bentley, The Snowflake Man.* Boyds, 2002. ISBN: 1-56397-689-7. Gr K-3

Bang, Molly. *Living Sunlight: How Plants Bring the Earth to Life.* The Blue Sky Press, 2009. ISBN: 978-0-545-044226. Gr PreS-3

Bauer, Marion Dane. *Wind.* Simon, 2003. ISBN: 0-689-85442-0. Gr K-3

Beck, W. H. *Glow: Animals with Their Own Night-Lights.* HMH Books for Young Readers, 2016. ISBN: 978-0-544-616666. Gr K-2

Berkes, Marianne. *Going around the Sun: Some Planetary Fun.* Dawn Publications, 2008. ISBN: 978-1-584-691006. Gr 1-7

Berne, Jennifer. *Manfish: A Story of Jacques Cousteau.* Chronicle Books, 2008. ISBN: 978-0-81186-063-5. Gr 1-4

Burleigh, Robert. *Solving the Puzzle under the Sea.* Simon & Schuster, 2016. ISBN: 978-1481-416-009. Gr PreS-3

Cassino, Mark. *The Story of Snow: The Science of Winter's Wonder.* Chronicle, 2009. ISBN: 978-0-8118-6866-2. Gr K-3

Cole, Joanna. *The Magic School Bus Lost in the Solar System.* Scholastic, 1992. ISBN: 0-590-41426-3. Gr K-3 Classic

Coleman, Janet Wyman. *Eight Dolphins of Katrina: A True Tale of Survival.* HMH Books for Young Readers, 2013. ISBN: 978-0-547-71923-8. Gr 1-4

Collins, Andrew. *Violent Weather: Thunderstorms, Tornadoes, and Hurricanes.* National Geographic Books, 2006. ISBN: 0-7922-5947-5. Gr K-3

Gall, Chris. *Awesome Dawson*. Little, Brown for Young Readers, 2013. ISBN: 978-0-316-213301. Gr PreS-2

Gans, Holly. *Let's Go Rock Collecting*. HarperCollins, 1997. ISBN: 978-0-064-451703. Gr K-4

Gray, Leon. *Giant Pacific Octopus: The World's Largest Octopus*. Bearport Publishing, 2013. ISBN: 978-1-617-72730-6. Gr 1-3

Guiberson, Brenda Z. *Frog Song*. Henry Holt and Co., 2013. ISBN: 978-0-805-09254-7. Gr PreS-3

HoPreSins, Lee Bennett. *Blast Off! Poems about Space*. HarperCollins, 1995. ISBN: 978-0-0602-4261-9. Gr 1-3

Kessler, Colleen. *100 Backyard Activities That Are the Dirtiest, Coolest, Creepy-Crawlist Ever*. Page Street Publishing, 2017. ISBN: 978-1624143731. Gr PreS-3

Kops, Deborah. *Exploring Space Robots*. Lerner, 2011. ISBN: 978-0-7613-5445-1. Gr K-3

Kudlinski, Kathleen W. *Boy Were We Wrong about the Solar System!* Dutton Juvenile, 2008. ISBN: 978-0-52546-979-7. Gr 1-3

Lawler, Janet. *Ocean Counting*. National Geographic Children's Books, 2013. ISBN: 978-1-426-31116-1. Gr PreS-1

Markle, Sandra. *Growing Up Wild: Penguins*. Atheneum, 2002. ISBN: 0-689-81887-4. Gr K-3

Markle, Sandra. *Outside and Inside Sharks*. Aladdin, 1999. ISBN: 978-0-6898-2683-2. Gr K-3

Martin, Jacqueline Briggs. *Snowflake Bentley*. Houghton Mifflin, 1998. ISBN: 978-0-39586-162-2. Gr K-3 Classic

Mizielinska, Aleksandra. *Under Water, Under Earth*. Big Picture Press, 2016. ISBN: 978-0-763-689223. Gr 1-3

Momatiuk, Yva. *Face to Face with Penguins*. National Geographic Children's Book, 2009. ISBN: 978-1-426-30561-0. Gr 2-5

Pattison, Darcy. *Nefertiti, the Spidernaut*. Mims House, 2016. ISBN: 978-1-629-440613. Gr K-6

Paul, Miranda. *One Plastic Bag: Isatou Ceesay and the Recycling Women of Gambia*. Millbrook Press, 2015. ISBN: 978-1-467-716086. Gr 1-4

Pfeffer, Wendy. *Life in a Coral Reef*. HarperCollins, 2009. ISBN: 978-0-06-029553-0. Gr K-3

Rockwell, Anne. *Clouds: Let's-Read-and-Find-Out Science*. Collins, 2008. ISBN: 978-0-06-445220-4. Gr PreS-1

Roy, Katherine. *Neighborhood Sharks: Hunting with the Great Whites of California's Farallon Islands*. David Macaulay Studio, 2014. ISBN: 978-1-59643-874-3. Gr 2-6

Rusch, Elizabeth. *Volcano Rising*. Charlesbridge, 2013. ISBN: 978-1-580-89408-1. Gr 1-4

Sayre, April Pulley. *Here Come the Humpbacks*. Charlesbridge, 2013. ISBN: 978-1-580-89406-7. Gr PreS-3

Sayre, April Pulley. *Raindrops Roll*. Beach Lane, 2015. ISBN: 978-1-48142-064-8. Gr K-2

Simon, Seymour. *Hurricanes*. Collins, 2007. ISBN: 978-0-061-170720. Gr K-4

Simon, Seymour. *The Moon*. Simon & Schuster Books for Young Readers; Revised Edition, 2015 (September 1, 2003). ISBN: 978-0-68983-563-6. Gr 2-6

Simon, Seymour. *Oceans*. Morrow, 2006. ISBN: 0-688-09453-8. Gr K-3

Sis, Peter. *Starry Messenger*. Farrar, Straus and Giroux, 1996. ISBN: 978-0-37447-027-2. Gr 1-5 Classic

Sisson, Stephanie Roth. *Star Stuff: Carl Sagan and the Mysteries of the Cosmos*. Roaring Brook, 2014. ISBN: 978-1-59643-960-3. Gr PreS-3

Stewart, Melissa. *Can an Aardvark Bark?* Beach Lane Book, 2017. ISBN: 978-1-481-458528. Gr PreS-3

Informational: Earth Science

Alonzo, Juan Carlos. *Land Mammals of the World: Notes, Drawings and Observations about Animals*. Walter Foster, 2017. ISBN: 978-1-633-222063. Gr 3-6

Aquilar, David A. *Seven Wonders of the Solar System*. Viking, 2017. ISBN: 978-0-451-476852. Gr 5 and Up

Berger, Lee Rand and Marc Aronson. *The Skull in the Rock: How a Scientist, a Boy, and Google Earth Opened a New Window on Human Origins*. National Geographic, 2012. ISBN: 978-1-4263-1053-9. Gr 5 and Up

Bonewitz, Ronald. *Nature Guides: Rocks and Minerals*. Doring Kindersley, 2012. ISBN: 978-0-756-690427. YA

Brown, Don. *The Great American Dust Bowl*. Houghton Harcourt, 2013. ISBN: 978-0-54781-550-3. Gr 2-6

Bruchac, Joseph. *Rachel Carson: Preserving a Sense of Wonder*. Fulcrum Publishing, 2009. ISBN: 978-1-555-916954. Gr 3-5

Burns, Loree Griffin. *Tracking Trash: Flotsam, Jetsam, and the Science of Ocean Motion.* Houghton, 2007. ISBN: 978-0-61858-131-3. Gr 6-9

Cerullo, Mary M. *Giant Squid.* Capstone Press, 2014. ISBN: 978-1-42967-541-3. Gr 4-7

Davis, Kenneth C. *Don't Know Much about the Universe: Everything You Need to Know about Outer Space but Never Learned.* Harper Paperbacks, 2004. ISBN: 978-0-060-932565. YA

Deem, James M. *Bodies from the Ice: Melting Glaciers and the Recovery of the Past.* Houghton Mifflin, 2008. ISBN: 978-0-6188-0045-2. Gr 5-8

DK. *The Rock and Gem Book.* DK Children, 2016. ISBN: 978-1-465-450708. Gr 3 and Up

DK Publishing. *Rock and Gem Book and Other Treasures of the Natural World.* Doring Kindersley, 2016. ISBN: 978-1465450708. Gr 3-7

Honovich, Nancy. *Ultimate Explorer Field Guide: Rocks and Minerals.* National Geographic Children's Books, 2016. ISBN: 978-1-426-323027. Gr 3-7

Hoyt, Erich. *Weird Sea Creatures.* Firefly Books, 2013. ISBN: 978-1-770-85191-7. Gr 5-12

Hughes, Catherine D. *National Geographic Little Kids First Book of the Ocean.* National Geographic Children's Books, 2013. ISBN: 978-1-426-31368-4. Gr PreS-3

Johnson, Ian Cullerton. *Seeds of Change: Planting a Path to Peace.* Lee & Low, 2010. ISBN: 978-1-600-603679. Gr 2-6

Johnson, Rebecca L. *Journey into the Deep: Discovering New Ocean Creatures.* Millbrook, 2010. ISBN: 978-0-7613-4148-2. Gr 4-6

Kudlinski, Kathleen W. *Boy Were We Wrong about the Solar System!* Dutton Jevenile, 2008. ISBN: 978-0-52546-979-7. Gr 1-3

Lourie, Peter. *Whaling Season: A Year in the Life of an Arctic Whale Scientist.* Houghton Mifflin, 2009. ISBN: 978-0-618-77709-9. Gr 4-8

Mills, Andrea. *True or False.* DK, 2014. ISBN: 978-1-46542-467-9. Gr 3-8

Monroe, Randall. *What If: Serious Scientific Answers to Absurd Hypothetical Questions.* HMH, 2014. ISBN: 978-0-54427-299-6. Gr 8 and Up

Murphy, Glenn. *How Loud Can You Burp? More Extremely Important Questions (and Answers!)* Flash Point Paper, 2009. ISBN: 978-1-59643-506-3. Gr 4-7 Reference

Rusch, Elizabeth. *Eruption: Volcanoes and the Science of Saving Lives.* Houghton Mifflin, 2013. ISBN: 978-0-81098-411-0. Gr 5-8

Rusch, Elizabeth. *The Next Wave: The Quest to Harness the Power of the Oceans*. HMH Books for Young Readers, 2014. ISBN: 978-0-544-099999. Gr 5-7

Sayre, April Pulley. *Raindrops Roll*. Beach Lane, 2015. ISBN: 978-1-48142-064-8. Gr K-2

Scott, Elaine. *Our Moon: New Discoveries about Earth's Closest Companion.*. Clarion Books, 2016. ISBN: 978-0-547-483948. Gr 5 and Up

Simon, Seymour. *Seymour Simon's Extreme Oceans*. Chronicle Books, 2013. ISBN: 978-1-452-10833-9. Gr 4-7

Smith, Miranda. *Rocks, Minerals and Gems*. Scholastic, 2016. ISBN: 978-0545947190. Gr 3-7

Sparrow, Giles. *Night Sky*. Scholastic, 2013. ISBN: 978-0-545-38374-5. Gr 3-7

Swinburne, Stephen R. *Sea Turtle Scientist*. Houghton Harcourt, 2014. ISBN: 978-054-7367-552. Gr 2-6

Thomas, Peggy. *Thomas Jefferson Grows a Nation*. Calkins Creek, 2015. ISBN: 978-1-620-916282. Gr 3-6

Tomecek, Steve. *National Geographic Kids Everything Rocks and Minerals*. National Geographic Children's Books, 2011. ISBN: 978-1426307683. Gr 3-7

Turner, Pamela S. *The Dolphins at Shark Bay*. Houghton Mifflin, 2013. ISBN: 978-0-54771-638-1. Gr 5-9

White, Roland. *Cleared for Takeoff*. Chronicle Books, 2016. ISBN: 978-1-452-135502. Gr 4-8

Wick, Walter. *A Drop of Water: A Book of Science and Wonder*. Scholastic, 1997. ISBN: 978-0-59022-197-9. Gr PreS-3 Classic

Yoder, Eric. *One Minute Mysteries: 65 More Short Mysteries You Solve with Science*. Platypus Media, 2012. ISBN: 978-1-938-49200-9-6. Gr 3-7

Zoehfeld, Kathleen Weidner. *National Geographic Readers: Rocks and Minerals*. National Geographic Children's Books, 2012. ISBN: 978-1-426-310186. Gr K-3

Physical Science: Definition

Any of the natural sciences (as physics, chemistry, and astronomy) that deal primarily with nonliving things.

—Merriam Webster

Picture Books: Physical Science

Adler, David A. *Things That Float and Things That Don't*. Holiday House, 2013. ISBN: 978-0-823-42862-5. Gr K-2

Berne, Jennifer. *On a Beam of Light: A Story of Albert Einstein*. Chronicle Books, 2013. ISBN: 978-0-81187-235-5. Gr 1-4

Boothroyd, Jennifer. *Give It a Push! Give It a Pull! A Look at Forces*. Lerner Classroom, 2010. ISBN: 978-0-761-360569. Gr 1-4

Chin, Jason. *Gravity*. Roaring Brook, 2014. ISBN: 978-1-59643-717-3. Gr K-2

Coan, Sharon. *Pushes and Pulls*. Teacher Created Materials, 2015. ISBN: 978-1-493-820528. Gr 1-2

Gerstein, Mordicai. *The Night World*. Little, Brown Books for Young Readers, 2015. ISBN: 978-0-316-188227. Gr PreS-3

Goldsmith, Mike. *Everything You Need to Know about Science*. Kingfisher, 2009. ISBN: 978-0-7534-6302-4. Gr K-3

Kolar, Bob. *Astroblast: Code Blue*. Scholastic, 2010. ISBN: 978-0-545-12104-0. Gr K-3

Kops, Deborah. *Exploring Space Robots*. Lerner, 2011. ISBN: 978-0-7613-5445-1. Gr K-3

Morgan, Emily. *Next Time You See a Sunset*. National Science Teachers Association, 2012. ISBN: 978-1-938-94626-4. Gr 1-3

Shields, Amy. *Little Kids First Big Book of Why*. National Geographic Books, 2011. ISBN: 978-1-4263-0793-5. Gr K-3

Simon, Seymour. *The Moon*. Simon & Schuster Books for Young Readers; Revised Edition, 2015 (September 1, 2003). ISBN: 978-0-68983-563-6. Gr 2-6

Sisson, Stephanie Roth. *Star Stuff: Carl Sagan and the Mysteries of the Cosmos*. Roaring Brook, 2014. ISBN: 978-1-59643-960-3. Gr PreS-3

Yoder, Eric. *One Minute Mysteries: 65 More Short Mysteries You Solve with Science*. Platypus Media, 2012. ISBN: 978-1-938-49200-9-6. Gr 3-7

Informational: Physical Science

Aldrin, Buzz. *Welcome to Mars*. National Geographic Children's Books, 2015. ISBN: 978-1-426-322068. Gr 3-7

Basher, Simon. *Basher Science: Extreme Physics*. Kingfisher, 2013. ISBN: 978-0-75346-956-9. Gr 5-9

Brallier, Jess. *Who Was Albert Einstein?* Turtleback Books, 2002. ISBN: 978-0-613-436526. Gr 3-5

Brown, Jordan D. *Crazy Concoctions: A Mad Scientist's Guide to Messy Mixtures*. Imagine, 2012. ISBN: 978-1-9361-4051-0. Gr 4-7

Chin, Jason. *Gravity.* Roaring Brook, 2014. ISBN: 978-1-59643-717-3. Gr K-2

Christensen, Bonnie. *I, Galileo.* Knopf Books for Young Readers, 2012. ISBN: 978-0-375-867538. Gr 3-7

Conkling, Winifred. *Radioactive! How Irene Curie and Lise Meitner Revolutionized Science and Changed the World.* Algonquin Young Readers, 2016. ISBN: 978-1-616-204150. Gr 5 and Up

Corey, Shana. *The Secret Subway.* National Geographic Children's Books, 2016. ISBN: 978-1-426-304620. Gr 5 and Up

Devorkin, David. *The Hubble Cosmos: 25 Years of New Vistas in Space.* National Geographic, 2015. ISBN: 978-1-426215575.

Dingle, Adrian. *How to Make a Universe with 92 Ingredients.* Owl Kids Books, 2013. ISBN: 978-0-7714-7008-7. Gr 4 and Up

Dinwiddie, Robert. *Nature Guide: Stars and Planets.* Doring Kindersley, 2012. ISBN: 978-0756690403. Reference

Gray, Theodore. *Elements: A Visual Exploration of Every Known Atom in the Universe.* Black Dog & Leventhal, 2009. ISBN: 978-1579128142. Reference

Gray, Theodore. *Molecules: The Elements and Architecture of Everything.* Black Dog & Leventhal, 2014. ISBN: 978-1579129712. Reference

Grove, Tim. *Milestones of Flight.* Harry N. Abrams, 2016. ISBN: 978-1-419-720031. Gr 5-7

Hawking, Stephen and Lucy Hawking. *George and the Big Bang.* Simon & Schuster, 2013. ISBN: 978-1-442-440067. Gr 3-7 Fiction/Big Bang Theory

Hawking, Stephen and Lucy Hawking. *George and the Unbreakable Code.* Simon & Schuster, 2016. ISBN: 978-1-481-466271. Gr 3-7 Fiction/Nature and How It Works

Hawking, Stephen and Lucy Hawking. *George's Cosmic Treasure Hunt.* Simon & Schuster, 2011. ISBN: 978-1-442-421752. Gr 3-7 Fiction/Space Travel

Hawking, Stephen and Lucy Hawking. *George's Secret Key to the Universe.* Simon & Schuster, 2009. ISBN: 978-1-416-985416. Gr 3-7 Fiction/Black Holes

Heinecke, Liz Lee. *Kitchen Science Lab for Kids: 52 Family Friendly Experiments from the Pantry.* Quarry Books, 2014. ISBN: 978-1592539253. Gr 2-5

Hollihan, Kerrie Logan. *Isaac Newton and Physics for Kids: His Life and Ideas with 21 Activities.* Chicago Review Press, 2009. ISBN: 978-1-556-527784. Gr 4-7

Homer, Holly. *The 101 Coolest Simple Science Experiments*. Page Street Publishing, 2016. ISBN: 978-1-624-141331. Gr 3-7

Hutchinson, Sam. *STEM Starters for Kids Science Activity Book: Packed with Activities and Science Facts*. Racehorse for Young Readers, 2017. ISBN: 978-1-631-581922. Gr 1-5

Kerrod, Robin. *The Way Science Works*. DK Children, 2002. ISBN: 978-078-9485-625. Gr PreS-12

Krull, Kathleen. *Lives of the Scientists: Experiments, Explosions (and What the Neighbors Thought)*. HMH Books for Young Readers, 2013. ISBN: 978-0-152-05909-5. Gr 4-7

Kudlinski, Kathleen W. *Boy Were We Wrong about the Solar System!* Dutton Jevenile, 2008. ISBN: 978-0-52546-979-7. Gr 1-3

Latham, Donna. *Bridges and Tunnels: Investigative Feats of Engineering with 25 Projects*. Nomad Press, 2012. ISBN: 978-1-936-749515. Gr 3-7

Mills, Andrea. *True or False*. DK, 2014. ISBN: 978-1-46542-467-9. Gr 3-8

Monroe, Randall. *What If: Serious Scientific Answers to Absurd Hypothetical Questions*. HMH, 2014. ISBN: 978-0-54427-299-6. Gr 8 and Up

Murphy, Glenn. *How Loud Can You Burp? More Extremely Important Questions (and Answers!)* Flash Point Paper, 2009. ISBN: 978-1-59643-506-3. Gr 4-7

O'Quinn, Amy M. *Marie Curie for Kids: Her Life and Scientific Discoveries, with 21 Activities and Experiments*. Chicago Review Press, 2016. ISBN: 978-1-613-733202. Gr 4 and Up

Panchyk, Richard. *Galileo for Kids: His Life and Ideas, 25 Activities for Kids*. Chicago Review, Press, 2005. ISBN: 978-1-556-525667. Gr 4 and Up

Pascal, Janet. *Who Was Isaac Newton?* Turtleback, 2014. ISBN: 978-0-606-361743. Gr 3-7

Pohlen, Jerome. *Albert Einstein and Relativity for Kids: His life and Ideas with 21 Activities*. Chicago Review Press, 2012. ISBN: 978-1-613-740286. Gr 4 and Up

Robinson, Tom. *The Everything Kid's Science Experiment Book: Boil Ice, Float Water*. Everything, 2001. ISBN: 978-1-580-625579. Gr 3-7

Rovelli, Carlo. *Seven Brief Lessons on Physics*. Riverhead Books, 2016. ISBN: 978-0-399-184413. YA

Sis, Peter. *Starry Messenger*. Farrar, Straus and Giroux, 1996. ISBN: 978-0-37447-027-2. Gr 1-5

Sparrow, Giles. *Night Sky*. Scholastic, 2013. ISBN: 978-0-545-38374-5. Gr 3-7

Wagner, Kathy. *The Everything Kids Astronomy Book*. Everything, 2008. ISBN: 978-1598695441. Gr 4-7

Yoder, Eric. *One Minute Mysteries: 65 More Short Mysteries You Solve with Science*. Platypus Media, 2012. ISBN: 978-1-938-49200-9-6. Gr 3-7

Series
Baby Animals (Carolrhoda)

Backyard Wildlife (Bellwether)

Big Cats (Abdo Kids)

Bookworms: Wonders of Nature (Marshall Cavendish)

Dinosaurs (Abdo Kids)

Even More Supersized! (Bearport Publishing)

Insects (Abdo Kids)

Let's-Read-and-Find-Out Science (Harper Collins)

National Geographic Little Kids First Big Books (National Geographic)

National Geographic Readers (National Geographic)

Once, in America (Apprentice Shop Books)

Reptiles (Abdo Kids)

The Science Behind (Capstone Press)

Science Tools (Capstone Press)

Scientists in the Field (HMH Books for Young Readers)

Sharks (Abdo Kids)

Spider (Abdo Kids)

Time for Kids Biographies (Harper Collins)

Watch Plants Grow (Gareth Stevens)

Why Science Matters (Heinemann-Raintree)

Chapter 7

TECHNOLOGY

Technology gives us power, but it does not and cannot tell us how to use that power. Thanks to technology, we can instantly communicate across the world, but it still doesn't help us know what to say.

—Jonathan Sacks
http://www.brainyquote.com

Technology includes the nature of technology and technology in the fields of medicine, agriculture, biotechnology, construction, manufacturing, transportation, communication, information, biomedical power, and energy. Technology is the modification of the natural world to meet human wants and needs, technology helps us to improve health, to grow and process food, to harness energy, to communicate more effectively, to process data faster and accurately, to move people and things easier, to make products enhance our lives and to build structures that provide shelter and comfort. The goal of technology is to make modifications in the world to meet human needs. It extends our abilities to change the world. It has to be made clear that technology is more than just computers!

—Dugger, William E. "STEM: Some Basic Definitions"
(Senior Fellow, International Technology and
Engineering Educators Association).
http://www.iteea.org

GENERAL QUESTIONS

What did you learn from this book?

What qualifications did the author have to write this book?

Where can you go to find out more information about this topic?

What was the purpose of this book?

What does the author want the reader to believe about this topic?

Was there anything in this book that you did not understand?

FEATURED AUTHORS AND ANNOTATIONS

Name: Kathryn Gibbs Davis

Place of Birth: Milwaukee, Wisconsin

About: Kathryn Gibbs Davis is the award-winning author of more than 20 fiction and nonfiction titles. These include picture books, early chapter books, middle grade novels, YA novels, and one film produced for children. She has also created two series (mystery and sports). Teachers appreciate Kathryn's fresh approach to history in two bestselling titles, *Wackiest White House Pets* (Parents' Choice Gold Award) and *First Kids* (2010 Oppenheim Gold Seal), which features Malia and Sasha Obama. In recent years, Kathryn has been invited back as guest author to several presidential libraries, including the Kennedy, Bush, Carter, Truman, and National First Ladies' Libraries.

Website: http://gibbsdavis.com

Annotated Title:

Davis, Kathryn Gibbs. *Mr. Ferris and His Wheel.* HMH Books for Young Readers, 2014. Gr K-3

> This is a biography about the American inventor George Ferris who invented the Ferris wheel. Ferris beat gravity and all sorts of problems to create this amusement and theme park delight. The book includes wonderful illustrations of the 1893 World's Fair where the Ferris wheel was first unveiled.

Name: Brian Floca

Place of Birth: Temple, Texas

About: Brian Floca is an American writer and illustrator of children's books. He graduated from Brown University, received his MFA from the School of Visual Arts, and now lives and works in Brooklyn, New York. Brian's books as author and illustrator include *Locomotive*, for which he received the 2014 Caldecott Medal; *Moonshot: The Flight of Apollo 11*; *Lightship*; *The Racecar Alphabet*; and *Five Trucks*. He has illustrated the Poppy Stories series, by Avi; *Ballet for Martha: Making Appalachian Spring*, by Jan Greenberg and Sandra Jordan; Kate Messner's Marty McGuire novels; and, most recently, Lynne Cox's *Elizabeth, Queen of the Seas*. In addition to the Caldecott, Brian's books have received four Robert F. Sibert Honor awards, an Orbis Pictus Award, an Orbis Pictus Honor, and a silver medal from the Society of Illustrators and have twice been selected for the *New York Times'* annual 10 Best Illustrated Books list.

Website: http://www.brianfloca.com/

Annotated Title:

Floca, Brian. *Locomotive.* Atheneum/Richard Jackson Books, 2013. ISBN: 978-1-41699-415-2.

> A Caldecott award winner, *Locomotive* begins with a family heading West in 1869 along with the crew that makes the train run. One can almost feel the motion, hear the sounds, and watch the scenery go by while traveling through the mountains to the end of the line: "Thanks to the locomotive, you've crossed the wide plains and desserts."

Name: Kathy Ceceri

Place of Birth: New York

About: Kathy Ceceri homeschooled her two sons from kindergarten until college and learned a lot in the process. As an educator, her goal has always been to keep things interesting for her students—and herself. She continues to do that through her articles for About.com, her hands-on activity books on topics like robotics and geography, and her classes and workshops for children and adults. In addition to her writing in print and on-line, Kathy leads STEAM programs at schools, libraries, museums, and Maker Faires around the country and presents workshops on homeschooling and hands-on learning to parents and classroom teachers. She is also a frequent guest on the Parenting Roundabout podcast.

Website: http://craftsforlearning.com/

Annotated Title:

Ceceri, Kathy. *Robotics: Discover the Science and Technology of the Future with 20 Projects.* Nomad Press: Build It Yourself, 2012. Gr 3-7

> This book provides a thorough and fascinating timeline of the history of robots. Ceceri tells everything you need to know about robots and how to make them. This book gives detailed instructions and lots of information and shows how to make robots out of ordinary craft materials and parts from old toys and household devices.

ACTIVITIES THAT CONNECT TECHNOLOGY

Robotics

Science	Technology	Engineering
How Nanorobots work—from *How Stuff Works* http://electronics .howstuffworks.com/ nanorobot.htm	Ceceri, Kathy. *Robotics: Discover the Science and Technology of the Future with 20 Projects*. Nomad Press: Build It Yourself, 2012. Gr 3-7	Build a robot out of discarded materials—http:// www.ziggityzoom.com/ activity/diy-recycled-robot -craft-project

Arts	Math
Create a robot out of art materials—http://fun-a -day.com/magnetic-robot -art-with-tin-cans/	Robot calculator game— http://www.playkidsgames .com/games/robot/ robotCalculator.htm

INVENTORS AND INVENTIONS

Bark, Jasper. *Journal of Inventions: Leonardo Da Vinci*. Silver Dolphin Books; Pop Edition, 2009. ISBN: 978-1-59223-908-5. Gr 2 and Up

Basher, Simon. *Basher Science: Technology A Byte-Sized World!* Kingfisher, 2012. ISBN: 978-0-75346-820-3. Gr 5-9

Bull, Peter. *Explorers: Robots*. Kingfisher, 2013. ISBN: 978-0-753-46816-6. Gr 2-5

Hubbard, Ben. *Hi Tech World: Cool Stuff.* Crabtree Publishing Company, 2010. ISBN: 978-0-77877-551-5. Gr 6 and Up

Oxlade, Chris. *The Camera.* Heinemann Library, 2010. ISBN: 978-0-43111-841-3. Gr 3-5

Salzmann, Mary Elizabeth. *Accordion to Zeppelin: Inventions from A to Z.* Super Sandcastle, 2008. ISBN: 1604530081. Gr 1 and Up

Spilsbury, Richard. *The Telephone.* Heinemann, 2010. ISBN: 978-1-43293-833-8. Gr 3-5

Ye, Ting-xing. *The Chinese Thought of It: Amazing Inventions and Innovations.* Annick Press, 2009. ISBN: 978-1-55451-195-2. Gr 5-7

BIBLIOGRAPHY

Picture Books

Adler, David A. *Simple Machines: Wheels, Levers, and Pulleys.* Holiday House, 2016. ISBN: 978-0-823-435722. Gr K-3

Barnett, Mac. *How This Book Was Made.* Disney-Hyperion, 2016. ISBN: 978-1-423-152200. Gr PreS-K

Barretta, Gene. *Now and Ben: The Modern Inventions of Benjamin Franklin.* Holt, 2006. ISBN: 978-0-80507-917-3. Gr K-4

Belloni, Giulia. *Anything Is Possible.* Owlkids, 2013. ISBN: 978-1-926-973913. Gr PreS-2

Bolte, Mari. *Amazing Story of the Combustion Engine: Max Axiom STEM Adventures.* Capstone Press, 2013. ISBN: 978-1-47653-103-8. Gr 2-4

Brown, Don. *Odd Boy Out: Young Albert Einstein.* HMH Books for Young Readers, 2008. ISBN: 978-0-54701-435-7. Gr PreS-3

Brown, Don. *A Wizard from the Start: The Incredible Boyhood and Amazing Inventions of Thomas Edison.* Houghton Mifflin, 2010. ISBN: 978-0-54719-487-5. Gr 2-6

Brown, Marc. *Arthur's Computer Disaster.* Little, Brown Books for Young Readers, 1999. ISBN: 978-0-316-105347. Gr PreS-3

Bryant, Jen. *Six Dots: A Story of Young Louis Braille.* Knopf Books for Young Readers, 2016. ISBN: 978-0-449-813379. Gr K-3

Burleigh, Robert. *Solving the Puzzle under the Sea.* Simon & Schuster, 2016. ISBN: 978-1481-416-009. Gr. PreS-3

Butterworth, Christine. *Where Did My Clothes Come From?* Candlewick, 2015. ISBN: 978-0-763-677503. Gr K-3

Collins, Suzanne. *When Charlie McButton Lost Power*. Puffin Books, 2007. ISBN: 978-0-142-408575. Gr PreS-3

Dahl, Michael. *Roll, Slope, Slide: A Book about Ramps*. Picture Window Books, 2006. ISBN: 978-1-404-819092. Gr K-4

Davis, Kathryn Gibbs. *Mr. Ferris and His Wheel*. HMH Books for Young Readers, 2014. ISBN: 978-054-7959-221. Gr K-3

DiPucchio, Kelly. *Clink*. Balzer + Bray, 2011. ISBN: 978-0-061-929281. Gr PreS-2

Dodds, Dayle Ann. *Henry's Amazing Machine*. Farrar, Straus and Giroux, 2004. ISBN: 978-0-374-329532. Gr PreS-3

Droyd, Ann. *Goodnight iPad*. Blue Rider Press, 2011. ISBN: 978-0-399-158568. Gr 1-2

Droyd, Ann. *If You Give a Mouse an iPhone*. Blue Rider Press, 2014. ISBN: 978-0-399-169267. Gr 1-2

Dyckman, Ame and Dan Yaccarino. *Boy and Bot*. Knopf Books for Young Readers, 2012. ISBN: 0375867562. Gr PreS-1

Eggers, Dave. *This Bridge Will Not Be Gray*. McSweeney's, 2015. ISBN: 978-1-940-450476. Gr PreS and Up

Enz, Tammy. *The Amazing Story of Cell Phone Technology: Max Axiom STEM Adventures*. Capstone Press, 2013. ISBN: 978-1-47650-137-6. Gr 3-4

Fern, Tracey. *Dare the Wind: The Record-Breaking Voyage of Eleanor Prentiss and the Flying Cloud*. Farrar, Straus and Giroux, 2014. ISBN: 978-037-4316-990. Gr K-3

Fleming, Candace. *Papa's Mechanical Fish*. Farrar, Straus and Giroux, 2013. ISBN: 978-0-374-39908-5. Gr K-3

Floca, Brian. *Locomotive*. Atheneum/Richard Jackson Books, 2013. ISBN: 141-6-99415-7. Gr PreS-5

Ford, Gilbert. *The Marvelous Thing That Came from a Spring: The Accidental Invention of the Toy That Swept the Nation*. Atheneum Books for Young Readers, 2016. ISBN: 978-1-481-450652. Gr PreS-3

Furstinger, Nancy. *Helper Robots*. Lerner Publishing Group, 2014. ISBN: 978-1-467-745086. Gr K-2

GrandPre, Mary. *Cleonardo, the Little Inventor*. Arthur A. Levine, 2016. ISBN: 978-0-439-357647. Gr K-4

Hadfield, Chris. *The Darkest Dark*. Little, Brown Books for Young Readers, 2016. ISBN: 978-0-316-394727. Gr PreS-3

Ito, Jeffrey. *Peter and Pablo The Printer: Adventures in Making the Future*. CreateSpace, 2016. ISBN: 978-1-539-168805. Gr 1 and Up

Kamkwamba, William. *The Boy Who Harnessed the Wind: Young Readers Edition*. Dial, 2012. ISBN: 978-0-8037-3511-8. Gr 1 and Up

Kraft, Betsy Harvey. *The Fantastic Ferris Wheel: The Story of Inventor George Ferris*. Henry Holt and Co., 2015. ISBN: 978-1-627-790727. Gr K-4

Krull, Kathleen. *The Boy Who Invented TV: The Story of Philo Farnsworth*. Knopf Books for Young Readers, 2009. ISBN: 978-0-37584-561-1. Gr 1-4

Kulling, Monica. *Spic-and-Span! Lillian Gilbreth's Wonder Kitchen*. Tundra Books, 2014. ISBN: 978-1-770-493803. Gr K-3

Lang, Heather. *Swimming with Sharks: The Daring Discoveries of Eugenie Clark*. Albert Whitman & Co., 2016. ISBN: 978-0-807-521878. Gr K-3

Liukas, Lindo. *Hello Ruby: Adventures in Coding*. Feiwel & Friends, 2015. ISBN: 978-1-250-065001. Gr PreS-3

Liukas, Lindo. *Hello Ruby: Journey inside the Computer*. Feiwel & Friends, 2017. ISBN: 978-1-250-065322. Gr PreS-3

Lowell, Barbara. *Daring Amelia*. Penguin Young Readers, 2016. ISBN: 978-0-448-487601. Gr 1-3

Lowell, Barbara. *George Ferris: What a Wheel*. Grosset & Dunlap, 2014. ISBN: 978-0-448-479262. Gr PreS-K

Lowery, Lawrence F. *Up, Up in a Balloon: I Wonder Why*. NSTA Kids, 2013. ISBN: 978-1-938-94614-1. Gr K-6

Masiello, Ralph. *Ralph Masiello's Robot Drawing Book*. Charlesbridge, 2011. ISBN: 978-1-57091-536-9. Gr K-3

Maurer, Tracey Nelson. *John Deere, That's Who*. Henry Holt and Co., 2017. ISBN: 978-1-627-791298. Gr PreS-3

McCarthy, Meghan. *Earmuffs for Everyone! How Chester Greenwood Became Known as the Inventor of Earmuffs*. S & D, 2015. ISBN: 978-1-48140-637-6. Gr K-3

McCarthy, Meghan. *Pop! The Invention of Bubble Gum*. Simon & Schuster, 2010. ISBN: 978-1-41697-970-8. Gr PreS-3

McCully, Emily Arnold. *Caroline's Comets: A True Story*. Holiday House, 2017. ISBN: 978-0-823-436644. Gr PreS-3

McCully, Emily Arnold. *Marvelous Mattie*. Farrar, Straus and Giroux, 2006. ISBN: 978-0-374-348106. Gr K-3

Meltzer, Brad. *I Am Amelia Earhart*. Dial Books, 2014. ISBN: 978-0-803-740822. Gr K-3

Monroe, Chris. *Monkey with a Tool Belt and the Noisy Problem*. Carolrhoda Books, 2009. ISBN: 978-0-822-592471. Gr K-3

Nelson, Robin. *From Wood to Baseball Bat*. Lerner Publications, 2015. ISBN: 978-1467-738910. Gr K-3

Norman, Kim. *The Bot That Scot Built*. Sterling Children's Books, 2016. ISBN: 978-1-454-910640. Gr PreS-2

Pattison, Darcy. *Burn: Michael Faraday's Candle*. Mims House, 2016. ISBN: 978-1-629-440446. Gr K-6

Pattison, Darcy. *Nefertit, the Spidernaut: The Jumping Spider Who Learned to Hunt in Space*. Mims House, 2013. ISBN: 978-1-629-440606. Gr K-6

Pratt, Mary K. *What Is Computer Coding?* Lerner Publications, 2015. ISBN: 978-1-467-783071. Gr 1-4

Robinson, Fiona. *Ada's Ideas: The Story of Ada Lovelace, the World's First Computer Programmer*. Abrams Books for Young Readers, 2016. ISBN: 978-1-419-718724. Gr 1-4

Schroeder, Alan. *Ben Franklin: His Wit and Wisdom from A to Z*. Holiday House, 2011. ISBN: 978-0-82341-950-0. Gr 1 and Up

Silverman, Buffy. *How Do Big Rigs Work?* Lerner Publications Group, 2016. ISBN: 978-1-467-795012. Gr K-2

Sis, Peter. *The Pilot and the Little Prince: The Life of Antoine de Saint-Exupery*. Farrar/Frances Forster Books, 2014. ISBN: 978-037-4380-694. Gr 2-6

Sisson, Stephanie Roth. *Star Stuff: Carl Sagan and the Mysteries of the Cosmos*. Roaring Brook Press, 2014. ISBN: 978-1-596-439603. Gr 1-3

Smith, Matthew Clark. *Lighter than Air: Sophie Blanchard, the First Woman Pilot*. Candlewick, 2017. ISBN: 978-0-763-677329. Gr 1-4

Stanley, Diane. *Ada Lovelace, Poet of Science*. Simon & Schuster, 2016. ISBN: 978-1-481-452496. Gr K-3

Thermes, Jennifer. *Charles Darwin's Around-the-World Adventure*. Abrams Books for Young Readers, 2016. ISBN; 978-1-419-721205. Gr K-3

Van Vleet, Carmella. *To the Stars! The First American Woman to Walk in Space*. Charlesbridge, 2016. ISBN: 978-1-580-896443. Gr K-3

Walker, Sally M. *Investigating Electricity*. Lerner, 2011. ISBN: 978-0-7613-5772-8. Gr K-3

Wallmark, Laurie. *Ada Byron Lovelace and the Thinking Machine*. Creston Books, 2015. ISBN: 978-1-939-547200. Gr K and Up

Whelan, Gloria. *Queen Victoria's Bathing Machine*. S & S, 2014. ISBN: 978-1-41692-753-2. Gr K-3

Williams, Karen Lynn. *Galimoto*. HarperCollins, 1991. ISBN: 978-0-688-109912. Gr PreS-3 Classic.

Winter, Jonah. *The Secret Project*. Beach Lane Books, 2017. ISBN: 978-1-481-469135. Gr K-3

Yaccarino, Dan. *Doug Unplugged*. Knopf Books for Young Readers, 2013. ISBN: 978-0-375-866432. Gr K-4

Informational

Arato, Rona. *Design It! The Ordinary Things We Use Every Day and the Not-So-Ordinary Ways They Came to Be*. Tundra Paper, 2010. ISBN: 978-0-88776-846-0. Gr 4-6

Barton, Chris. *Attack! Boss!! Cheat Code! A Gamer's Alphabet*. POW! 2014. ISBN: 978-1-57687-701-2. Gr 1-4

Barton, Chris. *Whoosh! Lonnie Johnson's Super-Soaking Stream of Inventions*. Charlesbridge, 2016. ISBN: 978-1-580-892971. Gr 2-5

Bascomb, Neal. *Sabotage: The Mission to Destroy Hitler's Atomic Bomb*. Arthur A. Levine, 2016. ISBN: 978-0-545-732437. Gr 8 and Up

Becker, Helaine. *National Geographic Kids Everything Space: Blast Off for a Universe of Photos, Facts, and Fun!* National Geographic Children's Books, 2015. ISBN: 978-1-426-320743. Gr 3-7

Becker, Helaine. *What's the Big Idea? Inventions That Changed Life on Earth Forever*. Maple Tree, 2009. ISBN: 978-1-8973-4960-1. Gr 3-6

Becker, Helaine. *Zoobots: Wild Robots Inspired by Real Animals*. Kids Can Press, 2016. ISBN: 978-1-554-539710. Gr 3-7

Benoit, Peter. *The Hindenburg Disaster*. Children's Press, 2011. ISBN: 978-0-531-20626-3. Gr 3-5

Berger, Lee R. *The Skull in the Rock: How a Scientist, a Boy, and Google Earth Opened a New Window on Human Origins*. National Geographic Books, 2012, ISBN: 978-1-42631-010-2. Gr 5 and Up

Blackburn, Ken. *Kids' Paper Air Plane Book*. Workman, 1996. ISBN: 978-0-7611-0478-0. Gr 2 and Up

Blumenthal, Karen. *Steve Jobs: The Man Who Thought Different*. Feiwel and Friends, 2012. ISBN: 978-1-2500-1557-0. Gr 7-10

Bow, James. *Maker Projects for Kids Who Love Robotics*. Crabtree, 2016. ISBN: 978-0-778-722663. Gr 4-7

Bridgman, Roger. *DK Eyewitness Books: Robot*. DK Children, 2004. ISBN: 978-0-75660-254-3. Gr 3-7

Briggs, Jason R. *Python for Kids: A Playful Introduction to Programming*. No Starch Press, 2012. ISBN: 978-1-5932-7407-8. Gr 5 and Up

Bruzzone, Catherine. *STEM Starters for Kids Technology Activity Book: Packed with Activities and Technology Facts*. Racehorse for Young Readers, 2016. ISBN: 978-1-631-581953. Gr 1-5

Buckley, James, Jr. *The Moon*. Penguin Young Readers, 2016. ISBN: 978-0-448-490212. Gr 3-4

Byrd, Robert. *Electric Ben: The Amazing Life and Times of Benjamin Franklin*. Dial, 2012. ISBN: 978-0-8037-3749-5. Gr 4-7

Callery, Sean. *Victor Wouk: The Father of the Hybrid Car*. Crabtree Publishing Company, 2009. ISBN: 978-0-7787-4664-5. Gr 4 and Up

Carson, Mary Kay. *Beyond the Solar System: Exploring Galaxies, Black Holes, Alien Planets, and More: A History with 21 Activities*. Chicago Review Press, 2013. ISBN: 978-1-613-74544-1. Gr 5-8

Carson, Mary Kay. *Inside Biosphere 2: Earth Science under Glass*. HMH Books for Young Readers, 2015. ISBN: 978-0-544-416642. Gr 5-7

Carson, Mary Kay. *Mission to Pluto: The First Visit to an Ice Dwarf and the Kuiper Belt*. HMH Books for Young Readers, 2017. ISBN: 978-0-544-416710. Gr 5-7

Casey, Susan. *Kids Inventing! A Handbook for Young Inventors*. Jossey-Bass, 2005. ISBN: 978-0-47-166-0-86-6. Gr 6 and Up

Cassidy, John. *The Klutz Book of Inventions*. Klutz, 2010. ISBN: 978-0-54561-114-5. Gr 3 and Up

Castaldo, Nancy. *The Story of Seeds*. HMH Books for Young Readers, 2016. ISBN: 978-0-54-320239. Gr 6 and Up

Ceceri, Kathy. *Computer*. DK Children, 2011. ISBN: 978-0-75668-265-1. Gr 3-7

Ceceri, Kathy. *Robotics: Discover the Science and Technology of the Future with 20 Projects (Build It Yourself)*. Nomad Press, 2012. ISBN: 978-1-93674-975-1. Gr 3-7

Cherrix, Amy. *Eye of the Storm: NASA, Drones, and the Race to Crack the Hurricane Code.* HMH Books for Young Readers, 2017. ISBN: 978-0-544-411654. Gr 5-7

Christensen, Victoria G. *How Batteries Work.* Lerner Publications, 2016. ISBN: 978-1-512-407815. Gr 3-7

Christensen, Victoria G. *How Conductors Work.* Lerner Publications, 2016. ISBN: 978-1-512-407822. Gr 3-7

Conkling, Winifred. *Radioactive! How Irene Curie and Lise Meitner Revolutionized Science and Changed the World.* Algonquin Young Readers, 2016. ISBN: 978-1-616-204150. YA

Cornell, Kari. *Minecraft Creator: Markus Notch Persson.* Lerner Classroom, 2016. ISBN: 978-1-467-707139. Gr 3-7

Cornell, Kari. *Urban Biologist Danielle Lee.* Lerner Classroom, 2016. ISBN: 978-1-467-797191. Gr 3-5

Davis, Joshua. *Spare Parts: Four Undocumented Teenagers, One Ugly Robot, and the Battle for the American Dream.* FSG Originals, 2014. ISBN: 978-0-374-534985. YA

Davis, Kathryn Gibbs. *Mr. Ferris and His Wheel.* Houghton Mifflin Harcourt, 2014. ISBN: 978-0-54795-922-1. Gr 2-6

Demuth, Patricia Brennan. *Who Is Bill Gates?* Grosset & Dunlap, 2013. ISBN: 978-0-448-463332. Gr 3-7

Diane, Carla. *Leo the Maker Prince: Journeys in 3-D Printing.* Maker Media Inc., 2013. ISBN: 978-1-4571-8314-0. Gr 3 and Up

DiPiazza, Francesca Davis. *Friend Me! 600 Years of Social Networking in America.* Twenty-First Century, 2012. ISBN: 978-0-7613-4607-4. Gr 5-8

DK Publishing. *3D Printing Projects.* DK, 2017. ISBN: 978-1-465-464767. Gr 4-7

DK Publishing. *Super Cool Tech.* DK Children, 2016. ISBN: 978-1-465-452054. Gr 3 and Up

Woodcock, Jon. *DK Workbooks: Computer Coding.* DK, 2014. ISBN: 978-1-46542-685-7. Gr 3-7

Eckerson, Nate. *Stopmotion Explosion: Animate Anything and Make Movies.* Nate Eckerson, 2011. ISBN: 978-0-98333-110-0. Gr 4 and Up

Firestone, Mary. *Nintendo: The Company and Its Founders.* Abdo, 2011. ISBN: 978-1-61714-809-5. YA

Fritz, Jean. *What's the Big Idea, Ben Franklin?* Putnam, 1976. ISBN: 978-0-698-20365-5. Gr 2-4 Classic

Garcia, Tracey J. *Thomas Edison*. Rosen Publishing Group, Inc., 2013. ISBN: 978-0-531-20949-3. Gr 5-8

Gigliotti, Jim. *Who Was George Washington Carver?* Grosset & Dunlap, 2015. ISBN: 978-0-448-483122. Gr 3-7

Glass, Andrew. *Flying Cars: The True Story*. Clarion Books, 2015. ISBN: 978-0-618-984824. Gr 5-8

Goldstein, Margaret J. *Astronauts*. Lerner Publications, 2017. ISBN: 978-1-512-425888. Gr 4-7

Goldstone, Lawrence. *Higher, Steeper, Faster: The Daredevils Who Conquered the Skies*. Little, Brown Books for Young Readers, 2017. Gr 4-8

Gonzales, Andrea. *Girl Code: Gaming, Going Viral, and Getting It Done*. Harper Collins, 2017. ISBN: 978-0-062-472502. YA

Grabham, Tim, Suridh Hassan, Dave Reeve, and Clare Richards. *Movie Maker: The Ultimate Guide to Making Films*. Candlewick, 2010. ISBN: 978-0-763-649494. Gr 3-7

Grabowski, John. *Television*. Lucent, 2011. ISBN: 978-1-4205-0169-8. YA

Graham, Ian. *Robot Technology*. Smart Apple, 2011. ISBN: 978-1-59920-533-5. YA

Gregory, Josh. *Race Cars: Science Technology Engineering*. Children's Press, 2015. ISBN: 978-0-53120-614-0. Gr 6-8

Greve, Meg. *The Internet*. Rourke, 2014. ISBN: 978-1-627-177658. Primary

Hamen, Susan E. *Google: The Company and Its Founders*. ABDO, 2011. ISBN: 978-1-61714-808-8. YA

Hammond, Richard. *Car Science*. DK Publishing, 2008. ISBN: 978-0-75664-026-2. Gr 2-7

Harbour, Jonathan. *Video Game Programming for Kids*. Cengage Learning, 2014. ISBN: 1305501829. Gr 3-7

Harrison, Geoffrey C. *Lethal Weapons*. Norwood House Press, 2013. ISBN: 978-1-599-535920. Gr 5-7

Hartland, Jessie. *Steve Jobs: Insanely Great*. Ember, 2016. ISBN: 978-0-307-982988. Gr 7 and Up

Heos, Bridget. *Blood, Bullets and Bones: The Story of Forensic Science from Sherlock to DNA*. Balzer + Bray, 2016. ISBN: 978-0-062-387622. Gr 8 and Up

Hile, Lori. *Getting Ahead: Drugs, Technology, and Competitive Advantage*. Heinemann, 2012. ISBN: 978-1-43295-978-4. Gr 6-8

Hillstrom, Laurie Collier. *Global Positioning Systems*. Lucent, 2011. ISBN: 978-1-4205-0325-8. YA

Hillstrom, Laurie Collier. *Ideas That Changed the World*. DK Children, 2013. ISBN: 978-146-5414-236. Gr 5-12

Hooks, Gwendolyn. *Tiny Stitches: The Life of Medical Pioneer Vivien Thomas*. Lee & Low, 2016. ISBN: 978-1-620-141564. Gr 2-5

Isaacs, Sally Senzell. *All about America: Stagecoaches and Railroads*. Kingfisher, 2012. ISBN: 978-0-7534-6516-5. Gr 4-6

Isogawa, Yoshihito. *The Lego Mindstorms EV3 Idea Book: 181 Simple Machines and Clever Contraptions*. No Starch Press, 2014. ISBN: 978-1-59327-532-7. Gr 3 and Up

Josefowicz, Chris. *Video Game Developer*. Gareth Stevens, 2010. ISBN: 978-1-4339-1958-9. Gr 4-6

Juettner, Bonnie. *The Large Hadron Collider*. Norwood House Press, 2013. ISBN: 978-1-603-575805. Gr 3-5

Kenney, Karen Latchana. *What Makes Medical Technology Safer?* Lerner Publications, 2015. ISBN: 978-1-467-779166. Gr 3-7

Kent, Peter. *Technology*. Kingfisher, 2009. ISBN: 978-0-7534-6307-9. Gr 4-7

Kopp, Megan. *Maker Projects for Kids Who Love Electronics (Be a Maker!)*. Crabtree, 2016. ISBN: 978-0-778-725817. Gr 3-7

Lanier, Troy and Clay Nichols. *Filmmaking for Teens: Pulling Off Your Shorts*. Michael Wiese Productions, 2010. ISBN: 978-1-932907-68-1. YA

The LEAD Project. *Super Scratch Programming Adventure: Learn to Program by Making Cool Games*. No Starch Press, 2012. ISBN: 978-1-59327-531-0. Gr 4 and Up

Lee, Dora. *Biomimicry: Inventions Inspired by Nature*. Kids Can, 2011. ISBN: 978-1-55453-467-8. Gr 3-6

Losure, Mary. *Issac the Alchemist: Secrets of Isaac Newton, Reveal'd*. Candlewick, 2017. ISBN: 978-0-763-670634. Gr 5 and Up

Macaulay, David. *The New Way Things Work*. HMH Books for Young Readers, 1998. ISBN: 978-039-5938-478. Gr 7 and Up

Marji, Majed. *Learn to Program with Scratch: A Visual Introduction to Programming with Games, Art, Science, and Math*. No Starch Press, 2014. ISBN: 978-1-59327-543-3. Gr 4 and Up

Marsico, Katie. *Tremendous Technology Inventions*. Lerner, 2013. ISBN: 978-1-46771-092-3. Gr 3-5

Maurer, Tracey Nelson. *John Deere, That's Who*. Henry Holt, 2017. ISBN: 978-1-627-791298. Gr PreS-3

McClafferty, Carla Kilough. *Tech Titans: One Frontier, Six Bios*. Scholastic Paper, 2012. ISBN: 978-0-5453-6577-2. Gr 5-7

Medina, Nico. *Who Was Jacques Cousteau?* Penguin Workshop, 2015. ISBN: 978-0-448-482347. Gr 3-7

Miller, Reagan. *Communication in the Ancient World*. Crabtree, 2011. ISBN: 978-0-7787-1733-1. Gr 5-7

Miller, Ron. *Seven Wonders of Space Technology*. Twenty-First Century Books, 2011. ISBN: 978-0-7613-5454-3. Gr 6-9

Mills, J. Elizabeth. *Creating Content: Maximizing Wikis, Widgets, Blogs, and More*. The Rosen Publishing Company, 2011. ISBN: 978-1-4488-1322-3. YA

Mitchell, Susan K. *Spy Tech: Digital Dangers*. Enslow, 2011. ISBN: 978-0-7660-3712-0. Gr 4-6

Mooney, Carla. *The Industrial Revolution: Investigate How Science and Technology Changed the World with 25 Projects*. Nomad, 2011. ISBN: 978-1-9363-1381-5. Gr 4-7

Mulder, Michelle. *Trash Talk: Moving toward a Zero-Waste World*. Orca Book Publishers, 2015. ISBN: 978-1-459-806924. Gr 3-7

Murphy, Jim. *Breakthrough! How Three People Saved "Blue Babies" and Changed Medicine Forever*. Clarion Books, 2015. ISBN: 978-0-547-821832. Gr 5-7

Murphy, Maggie. *High-Tech DIY Projects with 3-D Printing*. Powerkids Pr, 2014. ISBN: 978-1-47776-676-7. Gr 4 and Up

Murphy, Maggie. *High-Tech DIY Projects with Robotics*. Powerkids Pr, 2014. ISBN: 978-1-44776-675-0. Gr 5-8

Nardo, Don. *Destined for Space: Our Story of Exploration*. Capstone, 2012. ISBN: 978-1-4296-7540-6. Gr 3-5

Newquist, HP. *Abracadabra: The Story of Magic through the Ages*. Henry Holt and Co, 2015. ISBN: 978-0-312-593216. Gr 4-8

Older, Jules. *Snowmobile: Bombardier's Dream Machine*. Charlesbridge, 2012. ISBN: 978-1-58089-334-3. YA

Osborne, Dave. *Woodworking Projects with and for Children*. Dave's Shop Talk from DDFM Enterprises, 2014. Kindle

O'Shaughnessy, Tam. *Sally Ride: A Photobiography of America's Pioneering Women in Space*. Roaring Brook Press, 2015. ISBN: 978-1-596-439948. Gr 5-8

Oxlade, Chris. *Gaming Technology*. Smart Apple, 2011. ISBN: 978-1-59920-531-1. YA

Parker, Steve. *What about . . . Science and Technology*. Mason Crest, 2009. ISBN: 978-1-4222-1565-4. Gr 5-8

Pollack, Pam. *Who Was Steve Jobs?* Grosset & Dunlap, 2012. ISBN: 978-0-4484-7940-8. Gr 3 and Up

Price, Jane. *Underworld: Exploring the Secret World beneath Your Feet*. Kids Can Press, 2014. ISBN: 978-1-894-786898. Gr 3-7

Roland, James. *How Circuits Work*. Lerner Publications, 2016. ISBN: 978-1-512-407785. Gr 3-7

Rosenstock, Barb. *Ben Franklin's Big Splash: The Mostly True Story of His First Invention*. Calkins Creek, 2014. ISBN: 978-1-620-914465. Gr 3-7

Ross, Stewart. *Sports Technology*. Smart Apple, 2011. ISBN: 978-1-59920-534-2. YA

Rusch, Elizabeth. *Electrical Wizard: How Nikola Tesla Lit Up the World*. Candlewick, 2013. ISBN: 978-0-763-65855-7. Gr 2-5

Rusch, Elizabeth. *Mighty Mars Rovers: The Incredible Adventures of Spirit and Opportunity*. Houghton, 2012. ISBN: 978-0-5474-7881-4. Gr 5 and Up

Sande, Warren and Carter Sande. *Hello World! Computer Programming for Kids and Other Beginners*. Manning Publications, 2013. ISBN: 978-1-61729-092-3. YA

Scott, Elaine. *Our Moon: New Discoveries about Earth's Closest Companion*. Clarion Books, 2016. ISBN: 978-0-547-483948. Gr 4-8

Scott, Elaine. *Space, Stars, and the Beginning of Time: What the Hubble Telescope Saw*. Clarion Books, 2011. ISBN: 978-0-5472-4189-0. Gr 4 and Up

Shea, Therese. *Robotics Club: Teaming Up to Build Robots*. The Rosen Publishing Group, 2011. ISBN: 978-1-4488-1237-0. Gr 5 and Up

Sheinkin, Steve. *Bomb: Race to Build—and Steal—the World's Most Dangerous Weapon*. Flash Point, 2012. ISBN: 978-1-596-434875. Gr 5 and Up

Slade, Suzanne. *The Inventor's Secret: What Thomas Edison Told Henry Ford*. Charlesbridge, 2015. ISBN: 978-1-580-896-672. Gr 3-6

Sobey, Ed. *Electric Motor Experiments*. Enslow, 2011. ISBN: 978-0-766-033061. Gr 6-9

Sobey, Ed. *Robot Experiments*. Enslow, 2011. ISBN: 978-0-7660-3303-0. Gr 6-9

Spengler, Kremena T. *An Illustrated Timeline of Inventions and Inventors (Visual Timelines in History)*. Picture Window Books, 2011. ISBN: 978-1-40487-017-8. Gr 2-4

St. George, Judith. *So You Want To Be an Inventor?* Philomel, 2002. ISBN: 978-0-39923-593-1. Gr 3 and Up Classic

Steele, Philip. *Trains: The Slide-Out, See-Through Story of World-Famous Trains and Railroads.* Kingfisher, 2010. ISBN: 978-0-7534-6465-6. Gr K-6

Stine, Megan. *Where Is the Brooklyn Bridge?* Grosset & Dunlap, 2016. ISBN: 978-0-448-484242. Gr 3-7

Stone, Jerry. *One Small Step: Celebrating the First Men on the Moon.* Flash Point, 2009. ISBN: 978-1-5964-3491-0. Gr 1-5

Stone, Tanya Lee. *Almost Astronauts.* Candlewick, 2009. ISBN: 978-0-763-645021. Gr 5 and Up

Strom, Chris. *3D Game Programming for Kids: Create Interactive Worlds with Java Script.* Pragmatic Bookshelf, 2013. ISBN: 978-1-93778-544-4. Gr 3 and Up

Swanson, Jennifer. *Super Gear.* Charlesbridge, 2016. ISBN: 978-1-580-897204. Gr 6-9

Thomas, Peggy. *Thomas Jefferson Grows a Nation.* Calkins Creek, 2015. ISBN: 978-1-620-916282. Gr 3-6

Van Vleet, Carmella. *Explore Electricity with 25 Great Projects.* Nomad Press, 2013. ISBN: 978-1-61930-180-1. Gr K-4

Vance, Ashlee. *Elon Musk and the Quest for a Fantastic Future.* HarperCollins, 2017. ISBN: 978-0-062-463272. Gr 3 and Up

Venezia, Mike. *Steve Jobs and Steve Wozniak: Geek Heroes Who Put the Personal in Computers.* Scholastic, 2010. ISBN: 978-0-531-23730-4. Gr 2-4

Waxman, Laura Hamilton. *Aerospace Engineer Aprille Ericsson.* Lerner Publications, 2015. ISBN: 978-1-467-757935. Gr 3-6

Whale, David and Martin O'Hanlon. *Adventures in Minecraft.* Wiley, 2014. ISBN: 978-1-11894-691-6. YA

White, Roland. *Cleared for Takeoff.* Chronicle Books, 2016. ISBN: 978-1-452-135502. Gr 4-8

Woodcock, Jon. *Coding Games in Scratch: A Step-by-Step Visual Guide to Building Your Own Computer Games.* DK, 2016. ISBN: 978-1-465-439352. Gr 3-7

Woodcock, Jon. *Coding with Scratch Workbook.* DK, 2015. ISBN: 978-1-465-433922. Gr 1-4

Woodcock, Jon and Steve Setford. *Coding in Scratch Games Workbook.* DK, 2016. ISBN: 978-1-465-444820. Gr 1-4

Woods, Michael. *Ancients Machine Technology: From Wheels to Forges.* Twenty-First Century, 2011. ISBN: 978-0-7913-6523-5. YA

Woog, Adam. *You Tube.* Norwood House Press, 2009. ISBN: 978-1-59953-198-4. Gr 3 and Up

Wyckoff, Edwin Brit. *The Guy Who Invented Home Video Games: Ralph Baer and His Awesome Invention.* Enslow, 2010. ISBN: 978-0-7660-3450-1. Gr 3-5

Series

Awesome Inventions You Use Every Day (Lerner)

Blazers: See How It's Made (Capstone)

Build It Yourself (Nomad Press)

Connect with Electricity (Lerner)

Cool Careers: Cutting Edge (Gareth Stevens)

How Does Energy Work (Lerner)

How It Works (Wiley Blackwell)

Inventions That Changed the World (Heinemann)

Lightning Bolt Books: How Flight Works (Lerner)

Ordinary People Change the World (Penguin Group)

Robotics (Rosen)

Scientists in the Field Series (HMH)

Seven Wonders Series (Twenty-First Century)

Space Discovery Guides (Lerner)

Start to Finish, Second Series (Lerner Classroom)

STEM Adventures (Graphic Library)

STEM Trailblazer Bios (Lerner Classroom)

Tales of Invention (Heinemann)

Technology 360 (Lucent)

Technology in Ancient Cultures (Twenty-First Century)

Technology Pioneers (Abdo Kids)

Voices for Green Choices (Crabtree)

We Thought of It (Annick Press)

Who Was . . .? What Was . . .? Where Is . . .? (Grosset & Dunlap)

Chapter 8

ENGINEERING

Engineering stimulates the mind. Kids get bored easily. They have got to get out and get their hands dirty: make things, dismantle things, fix things. When the schools can offer that, you'll have an engineer for life.

—Bruce Dickinson
http://www.brainyquote.com

Engineering *includes aerospace, architecture, mining, chemical, electrical, environmental, industrial/systems, materials, ocean, mechanical, naval, research and development and design. It is the knowledge of the mathematical and natural sciences gained by study, experience, and practice applied with judgment and to develop ways to economically utilize the materials and forces of nature for the benefit of mankind. It is important for engineering and technology teachers to work together especially in the area of K-12 education.*
—Dugger, William E. "STEM: Some Basic Definitions"
(Senior Fellow, International Technology and
Engineering Educators Association).
http://www.iteea.org/

GENERAL QUESTIONS

What did you learn from this book?

What qualifications did the author have to write this book?

Where can you go to find out more information about this topic?

What was the purpose of this book?

What does the author want the reader to believe about this topic?

Was there anything in this book that you did not understand?

FEATURED AUTHORS AND ANNOTATIONS

Name: David Macaulay

Place of Birth: Lancashire, England

About: Macaulay's books have sold more than two million copies in the United States, have been translated into a dozen languages, and have been widely praised. *TIME* said of his work, "What [Macaulay] draws, he draws better than any other pen-and-ink illustrator in the world." His numerous awards include the Caldecott Medal, won for his book *Black and White*; the Boston Globe–Horn Book Award; an American Institute of Architects Medal; the Washington Children's Book Guild Nonfiction Award; and the Bradford Washburn Award, presented by the Museum of Science in Boston to an outstanding contributor to science.

Website: http://www.hmhbooks.com/davidmacaulay/

Annotated Title:

Macaulay, David. *Built to Last*. Houghton Mifflin, 2010. Gr 5 and Up
   This book is an updated edition of his original texts on castles, cathedrals, and mosques. Macaulay applied new information and research and created some new diagrams. Macaulay's work is amazing in beauty and detail.

Name: Stephen Biesty

Place of Birth: Coventry, United Kingdom

About: Stephen Biesty has worked as a freelance illustrator since 1985, creating a wide variety of information books for both adults and children. He studied illustration at Brighton Polytechnic gaining a BA and then went on to earn an MA at City of Birmingham Polytechnic where he specialized in historical and architectural cutaways. He lives in a small Somerset village with his wife and son. Stephen became internationally successful in the 1990s for his best-selling "Incredible Cross-Section" books published by Dorling Kindersley, which have sold over 3.5 million copies worldwide and have been printed in 16 languages. He has won several prestigious awards, including the *New York Times* Best Illustrated Book Award in 1993.

Website: http://www.stephenbiesty.co.uk/

Annotated Title:

Ross, Stewart. *Into the Unknown: How Great Explorers Found Their Way by Land, Sea and Air*. Illustrated by Stephen Biesty. Candlewick Press, 2011. ISBN: 978-0-76366-992-8. Gr 3-7
   The story of exploration is also the story of technological advances: fourteen journeys of Pytheas the Greek to most likely Iceland in 340BC to NASA's Apollo 11 mission to landing on the moon in 1969. Amazing foldouts, detailed cross-sections, and facts portray the journeys in this visual delight.

Name: Chris Oxlade

Place of Birth: England

About: Chris has written over 200 titles. Apart from writing books, he also worked as a rock-climbing instructor. Most of his spare hours were spent actually doing rock climbing, playing tennis, taking photographs, and trying to keep track of ever-changing technology. Much of his work involves explaining complex scientific ideas and modern technology clearly to the young reader. He has also written about sports, such as cricket, soccer, and climbing, and about photography.

Website: https://web.archive.org/web/20130317001302/ http://www.oxgan.pwp. blueyonder.co.uk/author/

Annotated Title:

Oxlade, Chris. *Scientriffic: Roller Coaster Science*. Silver Dolphin Books, 2014. Gr 1 and Up
This is a book and a box! The book has great illustrations and photos of roller coasters and many great experiments for learning about principles of physics like gravity, momentum, velocity, acceleration, friction, and centripetal force. Inside the box is a complete cardboard roller coaster to put together and try out using a marble!

ACTIVITIES THAT CONNECT ENGINEERING

Roller Coasters

Science	Technology	Engineering
How roller coasters work—http://www.kidzworld.com/ article/4633-how-roller -coasters-work	Build your own roller coaster—http:// discoverykids.com/games/ build-a-coaster/	Oxlade, Chris. *Scientriffic: Roller Coaster Science*. Silver Dolphin Books, 2014. Gr 1 and Up
Arts	**Math**	
How to draw a roller coaster. https:// www.youtube.com/watch? v=0Z1a3KxXXkE	Roller coaster designers need math—http:// www.educationworld.com/ a_lesson/TM/WS_vacation _coaster_facts.shtml	

INVENTORS AND INVENTIONS

Becker, Helaine. *What's the Big Idea? Inventions That Changed Life on Earth Forever*. Maple Tree Press, 2009. ISBN: 978-1-897-34961-8. Gr 2-7

Casey, Susan. *Kids Inventing! A Handbook for Young Inventors*. Jossey-Bass, 2005. ISBN: 978-0-47166-086-6. Gr 6 and Up

Eribach, Arlene. *The Kids' Invention Book*. Lerner Publications Group, 1999. ISBN: 978-0-82259-844-2. Gr 4 and Up

Opini, Bathsheba. *Africans Thought of It: Amazing Innovations*. Annick Press, 2011. ISBN: 978-1-55451-276-8. Gr 5-7

Raum, Elizabeth. *The History of the Computer*. Heinemann, 2007. ISBN: 978-1-40349-655-3. Gr 1 and Up

Spilsbury, Louise A. *The Airplane (Tales of Invention)*. Heinemann, 2010. ISBN: 978-1-43293-837-6. Gr 3-5

BIBLIOGRAPHY

Picture Books

Arnold, Tedd. *Green Wilma, Frog in Space*. Dial, 2009. ISBN: 978-0-8037-2698-7. Gr K-3

Barnett, Mac. *Oh No! (Or How My Science Project Destroyed the World)*. Hyperion, 2010. ISBN: 978-1-4231-2312-5. Gr K-3

Barretta, Gene. *Neo Leo: The Ageless Ideas of Leonardo da Vinci*. Henry Holt and Co., 2009. ISBN: 978-0-80508-703-1. Gr PreS-4

Barretta, Gene. *Now and Ben: The Modern Inventions of Benjamin Franklin*. Square Fish, 2008. ISBN: 978-0-31253-568-8. Gr K-4

Barretta, Gene. *Timeless Thomas: How Thomas Edison Changed Our Lives*. Square Fish, 2017. ISBN: 978-1-250-114785. Gr 1-5

Beaty, Andrea. *Iggy Peck, Architect*. Harry N. Abrams, 2007. ISBN: 978-0-81091-106-2. Gr PreS-3

Beaty, Andrea. *Rosie Revere, Engineer*. Harry N. Abrams, 2013. ISBN: 978-1-41970-845-9. Gr K and Up

Branley, Franklyn M. *Mission to Mars*. HarperCollins, 2002. ISBN: 978-0-064-452335. Gr 2-3

Breen, Steve. *Violet the Pilot*. Dial Books, 2008. ISBN: 978-0-803-731257. Gr PreS-3

Carle, Eric. *I See a Song*. Scholastic Trade, 1996. ISBN: 978-0-590-252133. Gr PreS-1

Coan, Sharon. *Pushes and Pulls*. Teacher Created Materials, 2015. ISBN: 978-1-493-820528. Gr 1-2

Dodds, Dayle Ann. *Henry's Amazing Machine*. Farrar, Straus and Giroux, 2004. ISBN: 0-374-32953-2. Gr K-3

Fleming, Candace. *Papa's Mechanical Fish*. Margaret Ferguson, 2013. ISBN: 978-0-374-39908-5. Gr 2-4

Floca, Brian. *Moonshot: The Flight of Apollo 11*. Atheneum, 2009. ISBN: 978-1-4169-5046-2. Gr K-3

Gerstein, Mordicai. *How to Bicycle to the Moon to Plant Sunflowers: A Simple but Brilliant Plan in 24 Easy Steps*. Roaring Brook Press, 2013. ISBN: 978-1-5964-3512-4. Gr PreS-2

Glover, David. *Pulleys and Gears*. Heinemann, 2006. ISBN: 1-4034-8564-X. Gr K-3

Grey, Mini. *Toys in Space*. Alfred A. Knopf, 2013. ISBN: 978-0-3079-7815-8. Gr K-2

Griffith, Victoria. *The Fabulous Flying Machines of Alberto Santos-Dumont*. Abrams, 2011. ISBN: 978-1-41970-011-8. Gr 2-6

Hale, Christy. *Dreaming Up: A Celebration of Building*. Lee & Low Books, 2012. ISBN: 978-1-600-606519. Gr K-3

Harrison, James. *Space*. Kingfisher, 2012. ISBN: 978-0-7534-6883-8. Gr K-3

Hayden, Kate. *Amazing Buildings*. DK, 2003. ISBN: 978-0-78949-220-3. Gr 1 and Up

Hughes, Catherine D. *National Geographic Little Kids First Big Book of Space*. National Geographic Books, 2012. ISBN: 978-1-4263-1014-0. Gr K-3

Jungman, Ann. *The Most Magnificent Mosque*. Frances Children's Books, 2007. ISBN: 978-1-8450-7085-4. Gr 1-4

Kolar, Bob. *Astroblast: Code Blue*. Scholastic, 2010. ISBN: 978-0-545-12104-0. Gr K-3

Kops, Deborah. *Exploring Space Robots*. Lerner, 2011. ISBN: 978-0-7613-5445-1. Gr K-3

Laden, Nina. *Roberto, the Insect Architect*. Chronicle Books, 2000. ISBN: 978-0-811-824561. Gr 1-4

Lord, Cynthia. *Hot Rod Hamster*. Scholastic Press, 2010. ISBN: 978-0-545-03530-9. Gr PreS-1

Low, William. *Machines Go to Work in the City*. Holt, 2012. ISBN: 978-0-8050-9050-5. Gr PreS-2

Lowery, Lawrence F. *Up, Up in a Balloon: I Wonder Why*. NSTA Kids, 2013. ISBN: 978-1-938946-14-1. Gr K-6

Macaulay, David and Sheila Keenan. *Castle: How It Works*. David Macaulay Studio, 2012 ISBN: 978-1-59643-766-1. Gr 2-3

Macaulay, David and Sheila Keenan. *Jet Plane: How It Works*. David Macaulay Studio, 2012. ISBN: 978-1-59643-767-8. Gr 2-3

Mezzanotte, Jim. *Giant Bulldozers*. Gareth, 2005. ISBN: 0-8368-4910-8. Gr K-3

Miyares, Daniel. *Float*. Simon & Schuster Books for Young Readers, 2015. ISBN: 978-1-481-415247. Gr PreS-3

Nelson, Kristin. *Farm Tractors*. Lerner Publications, 2003. ISBN: 978-0-822-506072. Gr K and Up

Nelson, Kristin. *From Tree to House*. Lerner Publications, 2004. ISBN: 978-0-822-513926. Gr K and Up Others

Norman, Kim. *The Bot That Scot Built*. Sterling Children's Books, 2016. ISBN: 978-1-454-910640. Gr PreS-2

O'Brien, Patrick. *The Hindenburg*. Holt, 2000. ISBN: 978-0-80506-415-8. Gr K-3

O'Brien, Patrick. *Steam, Smoke, and Steel: Back in Time with Trains*. Charlesbridge, 2000. ISBN: 0-88106-972-8. Gr K-3

O'Brien, Patrick. *You Are the First Kid on Mars*. Putnam Juvenile, 2009. ISBN: 978-0-3992-4634-0. Gr K and Up

O'Sullivan, Robyn. *The Wright Brothers Fly*. National Geographic Books, 2007. ISBN: 978-1-4263-0188-9. Gr K-3

Oxlade, Chris. *Flight*. Kingfisher, 2012. ISBN: 987-0-7534-6881-4. Gr K-3

Reinhart, Matthew. *Star Wars: A Galactic Pop-Up Adventure*. Scholastic, 2012. ISBN: 978-0-545-17616-3. Gr K-3

Reynolds, Paul A. *Going Places*. Atheneum Books for Young Readers, 2014. ISBN: 978-1-442-466081. Gr PreS-3

Rice, Dona Herweck. *Good Work: Simple Tools*. Teacher Created Materials, 2015. ISBN: 978-1-493-821419. Gr 1-2

Ritchie, Scott. *Look at That Building: A First Book of Structures*. Kids Can Press, 2011. ISBN: 978-1-554-536962. Gr PreS-2

Sis, Peter. *The Pilot and the Little Prince: The Life of Antoine de Saint-Exupery*. Farrar/Frances Forster Books, 2014. ISBN: 978-037-4380-694. Gr 2-6

Smith, Matthew Clark. *Lighter than Air: Sophie Blanchard, the First Woman Pilot*. Candlewick, 2017. ISBN: 978-0-763677329. Gr 1-4

Sobel, June. *B Is for Bulldozer: A Construction ABC*. Sandpiper, 2006. ISBN: 0-1520-2250-3. Gr K and Up

Spengler, Kremena. *An Illustrated Timeline of Inventions and Inventors*. Picture Window Books, 2011. ISBN: 978-1-404-870178. Gr 2-4

Spires, Ashley. *The Most Magnificent Thing*. Kids Can Press, 2014. ISBN: 978-1-554-537044. Gr PreS-2

Waldendorf, Kurt. *Hooray for Construction Workers*. Lerner Classroom, 2016. ISBN: 978-1-512-414738. Gr PreS-1

Waxman, Laura Hamilton. *Exploring the International Space Station*. Lerner, 2011. ISBN: 978-0-7613-5443-7. Gr K-3

Williams, Treat. *Air Show*. Hyperion, 2010. ISBN: 978-1-4231-1185-6. Gr PreS-2

Informational

Andrews, Beth L. *Hands-On Engineering: Real-World Projects for the Classroom*. Prufrock Press Inc., 2012. ISBN: 978-1-593-639228. Gr 4-7

Anthes, Emily. *Frankenstein's Cat: Cuddling up to Biotech's Brave New Beasts*. Farrar, Straus and Giroux, 2014. ISBN: 978-0-374-534240. YA

Barton, Chris. *Whoosh! Lonnie Johnson's Super-Soaking Stream of Inventions*. Charlesbridge, 2016. ISBN: 978-1-580-892971. Gr 2-5

Bascomb, Neal. *The New Cool: A Visionary Teacher, His FIRST Robotics Team, and the Ultimate Battle of Smarts*. Broadway Books, 2012. Gr 7 and Up

Basher, Simon. *Basher Basics: Space Exploration*. Kingfisher, 2013. ISBN: 978-0-75347-165-4. Gr 3-7

Basher, Simon. *Basher Science: Astronomy: Out of This World*. Kingfisher, 2009. ISBN: 978-0-75346-290-4. Gr 4-9

Beaty, Andrea. *Ada Twist's Big Project Book for Stellar Scientists*. Harry N. Abrams, 2017. ISBN: 978-1-419-730245 Gr K-2

Beaty, Andrea. *Iggy Peck's Big Project Book for Amazing Architects*. Abrams, 2017. ISBN: 978-1-683-351306. Gr K-2

Beaty, Andrea. *Rosie Revere's Big Project Book for Bold Engineers*. Abrams, 2017. ISBN: 978-1-613-125304. Gr K-2

Benson, Michael. *Beyond: A Solar System Voyage*. Abrams Books for Young Readers, 2009. ISBN: 978-0-8109-8322-9. Gr 3-9

Biesty, Stephen. *Stephen Biesty's Incredible Cross-Sections*. Dorling Kindersley Publishers Ltd, 1992. ISBN: 978-0-86218-807-7. Gr 3 and Up

Biskup, Agnieszka. *The Incredible Work of Engineers with Max Axiom, Super Scientist*. Capstone Press, 2013. ISBN: 978-1-62065-705-8. Gr 2-4

Bjorklund, Ruth. *Venus*. Marshall Cavendish, 2010. ISBN: 978-0-7614-4251-6. Gr 4-8

Bortz, Fred. *Wonders of Space Technology*. Twenty-First Century, 2011. ISBN: 978-0-7613-5453-6. YA

Brake, Mark. *Really, Really Big Questions about Space and Time*. Kingfisher, 2010. ISBN: 978-0-7534-6502-8. Gr 4-7

Brush, Jim. *Roller Coasters.* Sea-to-Sea, 2012. ISBN: 978-1-59771-329-0. Gr 3-7

Carmichael, L. E. *Amazing Feats of Civil Engineering.* Essential Library, 2014. ISBN: 978-1-624-034275. Gr 6 and Up

Carter, David. *The Elements of Pop-Up: A Pop-Up Book for Aspiring Paper Engineers.* Scholastic, 1999. ISBN: 0-689-82224-3. Gr 4-7

Casey, Susan. *Kids Inventing: A Handbook for Young Inventors.* Jossey-Bass, 2005. ISBN: 978-0-471-660866. Gr 6-12

Ceceri, Kathy. *Making Simple Robots: Exploring Cutting-Edge Robotics with Everyday Stuff.* Maker Media, 2015. ISBN: 978-1-457-183638. Gr 6-12

Ceceri, Kathy and Samuel Carbaugh. *Robotics: Discover the Science and Technology of the Future with 20 Projects.* Nomad, 2012. ISBN: 978-1-936-749751. Gr 3-7

Chaikin, Andrew. *Mission Control, This Is Apollo: The Story of the First Voyages to the Moon.* Penguin Group, 2009. ISBN: 978-0-6700-1156-8. Gr 6-9

Chin-Lee, Cynthia. *Amelia to Zora: Twenty-Six Women Who Changed the World.* Charlesbridge, 2008. ISBN: 978-1-57091-523-9. Gr 3-7

Cobb, Vicki. *Fireworks.* Millbrook Press, 2005. ISBN: 978-0-761-327714. Gr 4-8

Corey, Shana. *The Secret Subway.* National Geographic Children's Books, 2016. ISBN: 978-1-426-304620. Gr 5 and Up

Davis, Joshua. *Spare Parts: Four Undocumented Teenagers, One Ugly Robot, and the Battle for the American Dream.* FSG Originals, 2014. ISBN: 978-0-374-534981. Gr 7 and Up

DeCristofano, Carolyn Cinami. *A Black Hole Is Not a Hole.* Charlesbridge, 2012. ISBN: 978-1-5709-1783-7. Gr 5-8

Diamandis, Peter H. *Abundance: The Future Is Better Than You Think.* Free Press, 2012. ISBN: 978-1-451-614213. Gr 7 and Up

Dillon, Patrick. *The Story of Buildings: From the Pyramids to the Sydney Opera House and Beyond.* Candlewick, 2014. ISBN: 978-0-76366-990-4. Gr 4-7

Dingle, Adrian. *How to Make a Universe with 92 Ingredients.* Owl Kids Books, 2013. ISBN: 978-0-7714-7008-7. Gr 4 and Up

DK. *Find Out! Engineering.* DK, 2017. ISBN: 978-1-465-462343. Gr 1-4

Doeden, Matt. *Finding Out about Geothermal Energy.* Lerner Publishing Group, 2014. ISBN: 978-1-467-745543. Gr 2-5

Doeden, Matt. *Finding Out about Wind Energy.* Lerner Classroom, 2014. ISBN: 978-1-467-745581. Gr 2-5

Donovan, Sandra. *The Channel Tunnel.* Lerner Publications, 2003. ISBN: 978-0-82254-692-4. Gr 4 and Up

Doudna, Kelly. *The Kids Book of Simple Machines.* Mighty Media Kids, 2015. ISBN: 978-1-938-063596. Gr K-4

Ebner, Avivia. *Engineering Science Experiments.* Chelsea House, 2011. ISBN: 978-1-604-138528. Gr 4-8

Egan, Erin. *Hottest Race Cars.* Enslow, 2007. ISBN: 978-0-7660-2871-5. Gr 4-6

Elsmore, Warren. *Brick Wonders: Ancient, Modern, and Natural Wonders Made from Legos.* Baron's Educational Series, 2014. ISBN: 978-1-43800-411-2. YA

Enz, Tammy. *Build It: Invent New Structures and Contraptions.* Capstone Press, 2012. ISBN: 9781429679817. Gr 3-4

Fern, Tracey. *Dare the Wind: The Record-Breaking Voyage of Eleanor Prentiss and the Flying Cloud.* Farrar, Straus and Giroux, 2014. ISBN: 978-037-4316-990. Gr K-3

Finkelstein, Norman H. *Three Across: The Great Transatlantic Air Race of 1927.* Boyds Mills, 2008. ISBN: 978-1-5907-8462-4. Gr 5-8

Freedman, Russell. *The Wright Brothers: How They Invented the Airplane.* Holiday House, 1994. ISBN: 978-0-82341-082-8. Gr 5 and Up Classic

Friedman, Thomas L. *Hot, Flat, and Crowded: Why We Need a Green Revolution—And How It Can Renew America.* Picador, 2009. ISBN: 978-0-312-428921. Gr 7 and Up

Gifford, Clive. *Things That Go.* Kingfisher, 2011. ISBN: 978-0-7534-6593-6. Gr 3-5

Gigliotti, Jim. *Hottest Dragsters and Funny Cars.* Enslow Publishers, 2008. ISBN: 978-0-7660-2870-8. Gr 5-9

Goldfish, Meish. *Amazing Amusement Park Rides.* Bearport, 2011. ISBN: 978-1-6177-2304-9. Gr 1-5

Goldfish, Meish. *Fabulous Bridges.* Amicus, 2010. ISBN: 978-1-60753-132-6. Gr 5-7

Goldfish, Meish. *Spectacular Skyscrapers.* Bearport, 2011. ISBN: 978-1-6177-2303-2. Gr 1-5

Goldstein, Margaret J. *Astronauts: A Space Discovery Guide.* Lerner Publications, 2017. ISBN: 978-1-512-425888. Gr 4-7

Goodman, Susan E. *Ultimate Field Trip #5: Blasting Off to Space Academy.* Atheneum Books for Young Readers, 2001. ISBN: 978-0-689-830440. Gr 3-5

Graham, Ian. *Amazing Stadiums*. Amicus, 2010. ISBN: 978-1-60753-131-9. Gr 5-7

Gregory, Josh. *Race Cars: Science Technology Engineering*. Children's Press, 2015. ISBN: 978-0-53120-614-0. Gr 6-8

Gurstelle, William. *The Art of the Catapult: Build Greek Ballistae, Roman Onagers, English Trebuchets, and More Ancient Artillery*. Chicago Review Press, 2004. ISBN: 978-1-55652-526-1. Gr 4 and Up

Gutelle, Andrew. *Stock Car Kings*. Penguin Group, 2001. ISBN: 978-0-44842-4897. Gr 2-3

Hammond, Richard. *Car Science*. DK Publishing, 2008. ISBN: 978-0-75664-026-2. Gr 2-7

Hansem, Amy S. *Wind Energy: Blown Away*. Powerkids Press, 2010. ISBN: 978-1-435-897427. Gr 2-5

Heine, Florian. *13 Architects Children Should Know*. Prestel Junior, 2014. ISBN: 978-3-791-371849. Gr 3-7

Hewitt, Ben. *The Town That Food Saved*. Rodale Books, 2010. ISBN: 978-1-605-296869. Gr 7 and Up

Hickam, Homer H. *Rocket Boys*. Delta, 2000. ISBN: 978-0-385-333214. Gr 7 and Up

Holmes, Keith C. *Black Inventors: Crafting over 200 Years of Success*. Global Black Inventor Research Projects, Inc., 2008. ISBN: 978-0-979-957307. Gr 7 and Up

Holt, Nathalia. *Rise of the Rocket Girls: The Woman Who Propelled Us, from Missiles to the Moon to Mars*. Little, Brown and Company, 2016. ISBN: 978-0-316-338929. Gr 7 and Up

Isaacson, Walter. *Steve Jobs*. Simon & Schuster, 2015. ISBN: 978-1-501-127625. Gr 7 and Up

Jackson, Kay. *Navy Submarines in Action*. PowerKids Press, 2009. ISBN: 978-1-4358-2751-6. Gr 2 and Up

Jacoby, Jenny and Vicky Barker. *STEM Starters for Kids Engineering Activity Book: Packed with Activities and Engineering Facts*. Racehorse for Young Readers, 2017. ISBN: 978-1-631-581946. Gr 2-3

Johnson, Steven. *How We Got to Now: Six Innovations That Made the Modern World*. Riverhead Books, 2015. ISBN: 978-594-633935. Gr 7 and Up

Hughes, Catherine D. *National Geographic Little Kids First Big Book of Space*. National Geographic Books, 2012. ISBN: 978-1-4263-1014-0. Gr K-3

Hurley, Michael. *The World's Most Amazing Bridges*. Raintree, 2011. ISBN: 978-1-41094-249-4. Gr 3-5

Isogawa, Yoshihito. *The Lego Technic Idea Book: Simple Machines, Book I.* Starch Press, 2010. ISBN: 978-1-59327-277-7. Gr 4 and Up

Jacoby, Jenny. *STEM Starters for Kids Engineering Activity Book: Packed with Activities and Engineering Facts.* Racehorse for Young Readers, 2017. ISBN: 978-1-631-581946. Gr 1-5

Johnson, D. B. *Palazzo Inverso.* Houghton Mifflin Books for Children, 2010. ISBN: 978-0-54723-999-6. Gr PreS-2

Jones, Charlotte. *Mistakes That Worked.* Delacorte Books for Young People, 1994. ISBN: 978-0-38532-043-6. Gr 3-7

Kamkwamba, William. *The Boy Who Harnessed the Wind: Creating Currents of Electricity and Hope.* William Morrow, 2010. ISBN: 978-0-061-730337. Gr 6 and Up

Kenney, Karen Latchana. *What Makes Medical Technology Safer?* Lerner Publications, 2015. ISBN: 978-1-467-779166. Gr 3-7

Kenney, Karen Latchana. *What Makes Vehicles Safer?* Lerner Publications, 2015. ISBN: 978-1467-779135. Gr 3-7

Latham, Donna. *Bridges and Tunnels: Investigate Feats of Engineering with 25 Projects.* Nomad Press, 2012. ISBN: 978-1-936-749522. Gr 3-7

Levy, Matthys. *Engineering the City: How Infrastructure Works Projects and Principles for Beginners.* Paw Prints, 2008. ISBN: 978-1-43526-096-2. Gr 6 and Up

Macaulay, David. *Building Big.* HMH Books for Young Readers, 2004. ISBN: 978-0-618-465279. Gr 7 and Up

Macaulay, David. *Built to Last.* HMH Books for Young Readers, 2010. ISBN: 978-0-54734-240-5. Gr 5 and Up

Macaulay, David. *Castle.* HMH Books for Young Readers, 1982. ISBN: 978-0-39525-784-5. Gr 5-9 Classic

Macaulay, David. *Cathedral: The Story of Its Construction.* Sandpiper, 1981. ISBN: 978-0-39531-668-9. Gr 5-9 Classic

Macaulay, David. *Mill.* HMH Books for Young Readers, 1989. ISBN: 978-0-39552-019-2. Gr 5-9 Classic

Macaulay, David. *Mosque.* Houghton, 2003. ISBN: 978-0-61824-034-0. Gr 7 and Up

Macaulay, David. *Pyramid.* HMH Books for Young Readers, 1982. ISBN: 978-0-39532-121-8. Gr 5-9

Macaulay, David. *Unbuilding.* HMH Books for Young Readers, 1987. ISBN: 978-0-39545-425-1. Gr 5-9 Classic

Macaulay, David. *Underground*. HMH Books for Young Readers, 1983. ISBN: 978-0-39534-065-3. Gr 5-9 Classic

Mann, Elizabeth. *The Brooklyn Bridge: The Story of the World's Most Famous Bridge and the Remarkable Family That Built It*. Mikaya Press, 2006. ISBN: 978-1-93141-416-6. Gr 4-8

Mann, Elizabeth. *Empire State Building*. Mikaya, 2003. ISBN: 978-1-93141-406-7. Gr 4-8

Mann, Elizabeth. *The Hoover Dam: The Story of Hard Times, Tough People and the Taming of a Wild River*. Miyaka, 2006. ISBN: 978-1-93141-413-5. Gr 4-8

McCue, Camille. *Getting Started with Engineering: Think Like an Engineer! (Dummies Junior)*. For Dummies, 2016. ISBN: 978-1-119-291220. Gr 2-5

McCullough, David. *The Wright Brothers*. Simon & Schuster, 2016. ISBN: 978-1-476-728755. Gr 7 and Up

McKendry, Joe. *Beneath the Streets of Boston: Building America's First Subway*. Godine, 2005. ISBN: 1-56792-284-8. Gr 4-6

Miodownik, Mark. *Stuff Matters: Exploring the Marvelous Materials That Shape Our Man-Made World*. Mariner Books, 2015. ISBN: 978-0-544-483941. Gr 7 and Up

Mitchell, Don. *Driven: A Photobiography of Henry Ford*. National Geographic, 2010. ISBN: 978-1-4263-0155-1. Gr 4-7

Mooney, Carla. *Pilotless Planes*. Norwood, 2010. ISBN: 978-1-59953-381-0. Gr 5-7

Nardo, Don. *Destined for Space: Our Story of Exploration*. Capstone, 2012. ISBN: 978-1-4296-7540-6. Gr 3-5

Newman, Patricia. *Water Power*. Cherry Lake Publishing, 2013. ISBN: 978-1-610-808996. Gr 4-8

Ottaviani, Jim. *Fallout: J. Robert Oppenheimer, Leo Szilard, and the Political Science of the Atomic Bomb*. G. T. Labs, 2013. ISBN: 978-0-966-010-633. Gr 7 and Up

Owings, Lisa. *What Protects Us during Natural Disasters?* Lerner Publications, 2015. ISBN: 978-1-467-779142. Gr 3-7

Paris, Stephanie. *Engineering Feats and Failures*. Teacher Created Materials, 2012. ISBN: 978-1-433-348716. Gr 3 and Up

Paxman, Christine. *From Mud Huts to Skyscrapers*. Prestel Junior, 2012. ISBN: 978-3-791-371139. Gr 3-7

Platt, Richard. *Stephen Biesty's Incredible Everything*. DK Children, 1997. ISBN: 978-078-9420-497. Gr 4-6

Rand, Tom. *Kick the Fossil Fuel habit: 10 Clean Technologies to Save Our World.* Eco Ten Publishing, Inc., 2010. ISBN: 978-0-981-295206. OP

Reyes, Sandi. *Engineer through the Year: 20 Turnkey STEM Projects to Intrigue, Inspire and Challenge—Grades K-2.* SDE Crystal Springs Books, 2012. ASIN: 1935502379. Gr K-2

Reyes, Sandi. *Engineer through the Year: 20 Turnkey STEM Projects to Intrigue, Inspire and Challenge—Grades 3–5.* SDE Crystal Springs Books, 2012. ASIN: B00QM20SFC. Gr 3-5

Roberts, Dustyn. *Making Things Move: DIY Mechanisms for Inventors, Hobbyists, and Artists.* McGraw-Hill Education, 2010. ISBN: 978-0-071-741675. Gr 7 and Up

Ross, John F. *Enduring Courage: Ace Pilot Eddie Rickenbacker and the Dawn of the Age of Speed.* St. Martin's Press, 2014. ISBN: 978-1-25003-371-2. YA

Rubalcaba, Jill. *I. M. Pei: Architect of Time, Space, and Purpose.* Marshall Cavendish, 2011. ISBN: 978-0-7614-5973-6. YA

Rusch, Caroline Starr. *Mighty Mars Rovers: The Incredible Adventures of Spirit and Opportunity.* Houghton, 2012. ISBN: 978-0-5474-7881-4. Gr 5 and Up

Schyffert, Bea Uusma. *The Man Who Went to the Far Side of the Moon: The Story of Apollo 11 Astronaut Michael Collins.* Chronicle Books, 2003. ISBN: 0-8118-4007-7. Gr PreS-7

Shetterly, Margot Lee. *Hidden Figures: The American Dream and the Untold Story of the Black Women Mathematicians Who Helped Win the Space Race.* William Morrow Paperbacks, 2016. ISBN: 978-0-062-363602. Gr 7 and Up

Silverman, Buffy. *Simple Machines: Forces in Action.* Heinemann, 2016. ISBN: 978-1-484-636404. Gr 3-6

Sly, Alexandra. *Cars on Mars: Roving the Red Planet.* Charlesbridge, 2009. ISBN: 978-1-57091-462-1. Gr 5-7

Sobel, Dava. *Longitude: The True Story of a Lone Genius Who Solved the Greatest Scientific Problem of His Time.* Bloomsbury, 2007. OP

Sobey, Ed. *The Motorboat Book: Build and Launch 20 Jet Boats, Paddle Wheelers, Electric Submarines and More.* Chicago Review Press, 2013. ISBN: 978-1-6137-4447-5. Gr 4 and Up

Steele, Philip. *A City through Time.* DK Children, 2013. ISBN: 978-146-5402-493. Gr 2-5

Stewart, Ross. *Into the Unknown: How Great Explorers Found Their Way by Land, Sea and Air.* Illustrated by Stephen Biesty. Candlewick Press, 2011. ISBN: 978-0-76366-992-8. Gr 3-7

Sullivan, George. *Built to Last: Building America's Amazing Bridges, Dams, Tunnels, and Skyscrapers.* Scholastic Nonfiction, 2005. ISBN: 0-439-51737-0. Gr 4-7

Swanson, Jennifer. *How Hybrid Cars Work*. Child's World, 2011. ISBN: 978-1-6097-3217-2. Gr 4-6

Tailfeather, Speck Lee. *Architecture According to Pigeons*. Phaidon Press, 2013. ISBN: 978-0-714-863894. Gr 2-6

Thimmesh, Catherine. *Girls Think of Everything: Stories of Ingenious Inventions by Women*. HMH Books for Young Readers, 2002. ISBN: 978-0-61819-563-3. Gr 4-6

VanCleave, Janice. *Janice VanCleave's Engineering for Every Kid: Easy Activities That Make Learning Science Fun*. Wiley, 2007. ISBN: 978-047-1471-182-0. Gr 3 and Up

Verstraete, Larry. *Surviving the Hindenburg*. Sleeping Bear, 2012. ISBN: 978-1-5836-6787-0. Gr 1-4

Weitzman, David. *Skywalkers: Mohawk Ironworkers Build the City*. Flash Point, 2010. ISBN: 978-1-59643-162-1. Gr 4-7

Willardson, Ben. *Exploring Careers with Kids: ABCs of Civil Engineering*. CreateSpace Independent Publishing Platform, 2016. ISBN: 978-1-532-873249. Gr PreS-K

Woodroffe, David. *Making Paper Airplanes: Make Your Own Aircraft and Watch Them Fly!* Skyhorse Publishing, 2012. ISBN: 978-1-62087-168-3. Gr K-3

Zaunders, Bo. *Feathers, Flaps, and Flops: Fabulous Early Fliers*. Dutton, 2001. ISBN: 0-525-46466-2. Gr 4-6

Zimmermann, Karl. *Ocean Liners: Crossing and Cruising the Seven Seas*. Boyds Mills Press, 2008. ISBN: 978-1-59078-552-2. Gr 4 and Up

Series

Build It Yourself (Rakuten Overdrive)

Connect with Electricity (Lerner Books)

Construction Machines Series (Abdo Kids)

Engineering Keeps Us Safe (Lerner)

Giant Vehicles (Gareth)

Graphic Science and Engineering in Action (Capstone)

Great Achievements in Engineering (ABDO Publishing)

Great Building Feats (Lerner Publications)

Hooray for Community Helpers (Lerner Publishing)

I Wonder Why (NSTA Kids)

Machines on the Farm (Abdo Kids)

Mighty Machines (Capstone)

Simple Machines (Heinemann)

Space (Marshall Cavendish)

Start to Finish (Lerner Classroom)

Stem Trailblazer Bios (Lerner Classroom)

Trucks (Abdo Kids)

U.S. Armed Forces Series (Abdo Kids)

Wild Wheels (Enslow)

Chapter 9

ARTS

Science provides an understanding of a universal experience. Arts provide a universal understanding of a personal experience.

—Mae Jemison
http://www.brainyquote.com

Arts *includes sociology, education, philosophy, psychology, history, language, sharing knowledge with language arts, a working knowledge of manual and physical arts, better understanding of the past and present through fine arts, and understanding developments with social and liberal arts.*
—Dugger, William E. "STEM: Some Basic Definitions"
(Senior Fellow, International Technology and
Engineering Educators Association).
http://www.iteea.org

GENERAL QUESTIONS

What did you learn from this book?

What qualifications did the author have to write this book?

Where can you go to find out more information about this topic?

What was the purpose of this book?

What does the author want the reader to believe about this topic?

Was there anything in this book that you did not understand?

FEATURED AUTHORS AND ANNOTATIONS

Name: Robert Burleigh

Place of Birth: Chicago, Illinois

About: Robert Burleigh has written many books for children from preschool through middle school. He has written biographies and survival stories. Robert Burleigh is an award-winning author. His titles for children include *The Adventures of Mark Twain by Huckleberry Finn*, illustrated by Barry Blitt; *Night Flight*, illustrated by Wendell Minor; and *Black Whiteness*, illustrated by Walter Lyon Krudop. His many other books include *Hoops*; *Stealing Home*; and *Clang! Clang! Beep! Beep!* He lives in Michigan.

Website: http://www.robertburleigh.com

Annotated Title:

Burleigh, Robert. *George Bellows: Painter with a Punch*. Harry N. Abrams, 2012. ISBN: 978-1-41970-166-5.

> Born in 1882, in Columbus, Ohio, George Bellows played basketball and baseball and was an artist. At 22, he left home and moved to New York to fulfill his goal of becoming a great artist. He died of appendicitis at the age of 42. The *New York Times* stated in his obituary that he "died before his work was done." This book includes a list of reproductions of his paintings and locations of where his work can be seen.

Name: Lois Ehlert

Place of Birth: Beaver Dam, Wisconsin

About: Lois Ehlert has created numerous inventive, celebrated, and best-selling picture books, including *Chicka Chicka Boom Boom*, *The Scraps Book*, *Mice*, *Ten Little Caterpillars*, *Rrralph!*, *Lots of Spots*, *Boo to You!*, *Leaf Man*, *Waiting for Wings*, *Planting a Rainbow*, *Growing Vegetable Soup*, and *Color Zoo*, which received a Caldecott Honor. Lois Ehlert is an author and illustrator of children's books, most having to do with nature. Ehlert won the Caldecott Honor for *Color Zoo*. She lives in Milwaukee, Wisconsin.

Website: http://authors.simonandschuster.com/Lois-Ehlert/1877089

Annotated Title:

Ehlert, Lois. *The Scraps Book: Notes from a Colorful Life*. Beach Lane Books, 2014. ISBN: 978-1-44243-571-1.

> In a compilation of notes from a colorful life, Lois Ehlert's parents made things with their hands and shared the scraps, tools, materials, and space for her to create and dream. Lois explains where she gets her ideas and how she figures out what should go on each page and works until the words and pictures together tell a story.

Name: Jeannette Winter

Place of Birth: Chicago, Illinois

About: Jeanette Winter is the acclaimed author/illustrator of many highly regarded picture books, including *The Librarian of Basra: A True Story from Iraq*; *Mama: A True Story in Which a Baby Hippo Loses His Mama During a Tsunami, but Finds a New Home, and a New Mama*; *Wangari's Trees of Peace: A True Story from Africa*; *Nasreen's Secret School: A True Story from Afghanistan*; *Biblioburro: A True Story from Colombia*; *Henri's Scissors*; *Mr. Cornell's Dream Boxes*; and most recently, *Malala, A Brave Girl from*

Pakistan/Iqbal, A Brave Boy from Pakistan. She lives with her husband, artist Roger Winter, in New York.

Website: http://authors.simonandschuster.com/Jeanette-Winter/64041479

Annotated Title:

Winter, Jeanette. *Malala, A Brave Girl from Pakistan/Iqbal, A Brave Boy from Pakistan*. Beach Lane, 2014. Gr 1-5

> Beautifully colorful artwork enhances the story of Malala of Pakistan who spoke out in favor of girls attending school and was shot but survived to continue her work. Malala won the Nobel Peace Prize in 2014. Iqbal, a Pakistani boy, also shot for speaking out against child slavery in Pakistan is also featured in this book. Iqbal did not survive his wounds.

ACTIVITIES THAT CONNECT ARTS

Collages

Science	Technology	Engineering
Nature collage—http://kidsactivitiesblog.com/14525/nature-craft-collage	Create a collage using the computer—http://www.technokids.com/blog/apps/digital-collage-in-the-classroom/	How to make a 3-D collage—http://victoriarestrepo.com/2013/02/21/art-project-for-kids-how-to-make-a-3d-collage/

Arts	Math
Ehlert, Lois. *The Scraps Book: Notes from a Colorful Life*. Beach Lane, 2014. Gr K-5	Place Value Collage—https://vimeo.com/90590246

INVENTORS AND INVENTIONS

Arato, Rona. *Design It: The Ordinary Things We Use Every Day and Not-So-Ordinary Ways They Came to Be*. Tundra Books, 2010. ISBN: 978-0-88776-846-0. Gr 4-7

Bampton, Claire. *The Journal and Historical Record of Invention and Discovery*. Arcturus Publishing Limited, 2014. ISBN: 978-1-78212-096-4. Gr 5 and Up

Hopkins, Lee Bennett. *Incredible Inventions*. Greenwillow, 2009. ISBN: 978-0-06087-245-8. Gr 1-5

Thompson, Ruth. *The Science and Inventions Creativity Book: Games, Models to Make, High-Tech Craft Paper, Stickers, and Stencils*. Barron's Educational Series, 2013. ISBN: 978-1-438-00251-4. Gr 1-6

Van Vleet, Carmelia. *Amazing Ben Franklin Inventions You Can Build Yourself*. Nomad Press, 2007. ISBN: 978-0-977-12947-8. Gr 4-7

BIBLIOGRAPHY

Picture Books

Ackerman, Karen. *Song and Dance Man*. Alfred A. Knopf, 1988. ISBN: 978-0-394-893303. Gr PreS-2 Classic

Andrews, Troy. *Trombone Shorty*. Harry N. Abrams, 2015. ISBN: 978-1-419-714658. Gr PreS-3

Applegate, Katherine. *The One and Only Ivan*. Harper, 2012. ISBN: 978-0-06-199225-4. Gr 3-7

Austin, Mike. *Monsters Love Colors*. HarperCollins, 2013. ISBN: 978-0-06212-594-1. Gr PreS-3

Barnett, Mac. *Extra Yarn*. Balzer + Bray, 2012. ISBN: 978-0-061-953385. Gr PreS-3

Bates, Katharine Lee. *America the Beautiful*. Orchard Books, 2013. ISBN: 978-0-5454 -9207-2. Gr K-3

Beaty, Andrea. *Iggy Peck, Architect*. Harry N. Abrams, 2007. ISBN: 978-0-8109-1106-2. Gr K-3

Beaty, Andrea. *Artist Ted*. Simon & Schuster, 2012. ISBN: 978-1-4169-5374-6. Gr PreS-2

Benson, Kathleen. *Draw What You See: The Life and Art of Benny Andrews*. Clarion Books, 2015. ISBN: 978-0-544-104877. Gr 2-4

Bliss, Harry. *Bailey at the Museum*. Scholastic, 2012. ISBN: 978-0-545-23345-3. Gr PreS-K

Brown, Monica. *My Name Is Celia: The Life of Celia Cruz*. Cooper Square Publishing, 2004. ISBN: 978-0-873-588720. Gr K-3

Brown, Monica. *Tito Puente, Mambo King*. Rayo, 2013. ISBN: 978-0-061-227837. Gr PreS-3

Bryant, Jen. *A Splash of Red: The Life and Art of Horace Pippin*. Knopf Books for Young Readers, 2013. ISBN: 978-0-3758-6712-5. Gr K-3

Burleigh, Robert. *Edward Hopper Paints His World*. Holt, 2014. ISBN: 978-080-507 -87-529. Gr 2-6

Burleigh, Robert. *George Bellows: Painter with a Punch*. Abrams, 2012. ISBN: 978-1-4197 -0166-5. Gr 3-7.

Campoy, F. Isabel. *Maybe Something Beautiful: How Art Transformed a Neighborhood.* HMH Books for Young Readers, 2016. ISBN: 978-0-544-357693. Gr PreS-2

Cline-Ransome, Lesa. *Benny Goodman and Teddy Wilson: Taking the Stage as the First Black-and-White Jazz Band in History.* Holiday House, 2014. ISBN: 978-0-823-423620. Gr 2-6

Cline-Ransome, Lesa. *Just a Lucky So and So: The Story of Louis Armstrong.* Holiday House, 2016. ISBN: 978-0-823-434282. Gr PreS-3

Cline-Ransome, Lesa. *My Story, My Dance: Robert Battle's Journey to Alvin Ailey.* Simon & Schuster, 2015. ISBN: 978-1-481-422215. Gr 2-5

Colon, Raul. *Draw.* S & S, 2014. ISBN: 978-144-2494-923. Gr 1-3 Wordless

Daly, Cathleen. *Emily's Blue Period.* Roaring Brook, 2014. ISBN: 978-159-6434-691. Gr 1-3

Davies, Jacqueline. *The Boy Who Drew Birds: A Story of John James Audubon.* Houghton Mifflin Books for Children, 2004. ISBN: 978-0-6182-4343-3. Gr K-3

Dempsey, Kristy. *A Dance Like Starlight: One Ballerina's Dream.* Philomel, 2014. ISBN: 978-039-9252-846. Gr 2-6

DePaola, Tomie. *The Art Lesson.* Puffin, 1997. ISBN: 978-0-7807-4073-0. Gr 1 and Up Classic

Dillon, Leo and Diane Dillon. *Jazz on a Saturday Night.* The Blue Sky Press, 2007. ISBN: 978-0-590-478939. Gr PreS-3

Durango, Julia. *Under the Mambo Moon.* Charlesbridge, 2011. ISBN: 978-1-570-917233. Gr 3 and Up

Ehrhardt, Karen. *This Jazz Man.* Harcourt Children's Books, 2006. ISBN: 978-0-1529 -5307-9. Gr 1-5

Engle, Margarita. *Drum Dream Girl: How One Girl's Courage Changed Music.* HMH Books for Young Readers, 2015. ISBN: 978-0-544-102293. Gr PreS-3

Engle, Margarita. *The Sky Painter.* Two Lions, 2015. ISBN: 978-1-477-826331. Gr 1-3

Falken, Linda. *Can You Find It? America.* Abrams Books for Young Readers, 2010. ISBN: 978-1-58839-334-0. Gr 2-5

Garriel, Barbara. *I Know a Shy Fellow Who Swallowed a Cello.* Boyds Mills Press, 2004. ISBN: 978-1-5907-8043-5. Gr K-2

Gollub, Matthew. *The Jazz Fly (Book with Audio CD)*. Tortuga Press, 2000. ISBN: 978-1-8899-1017-8. Gr 1 and Up

Greenberg, Jan. *Ballet for Martha: Making Appalachian Spring*. Flash Point, 2010. ISBN: 978-1-5964-3338-8. Gr 2-6

Guarnaccia, Steven. *The Three Little Pigs: An Architectural Tale*. Abrams Books for Young Readers, 2010. ISBN: 978-0-8109-8941-2. Gr 1 and Up

Hall, Michael. *A Perfect Square*. Greenwillow Books, 2011. ISBN: 978-0-061-915130. Gr PreS-3

Hall, Michael. *Red: A Crayon's Story*. Greenwillow Books, 2015. ISBN: 978-0-062-252074. Gr PreS-3

Hawkes, Kevin. *Remy and Lulu*. Knopf, 2014. ISBN: 978-044-9810-873. Gr 1-3

Herkert, Barbara. *Sewing Stories: Harriet Powers' Journey from Slave to Artist*. Knopf Books for Young Readers, 2015. ISBN: 978-0-385-754620. Gr K-3

Hill, Laban Carrick. *When the Beat Was Born*. Roaring Brook Press, 2013. ISBN: 978-1-596-43540. Gr 1-5

Holm, Jennifer L. *Babymouse: The Musical*. Random House Books for Young Readers, 2009. ISBN: 978-0-375-843884. Gr 2-5

Holm, Jennifer L. *Babymouse: Rock Star*. Random House Books for Young Readers, 2006. ISBN: 978-0-375-832321. Gr 2-5

Hood, Susan. *Ada's Violin: The Story of the Recycled Orchestra of Paraguay*. Simon & Schuster Books for Young Readers, 2016. ISBN: 978-1-481-430951. Gr PreS-3

Idle, Molly. *Flora and the Flamingo*. Chronicle Books, 2013. ISBN: 978-1-452-110066. Gr PreS-1

Idle, Molly. *Flora and the Peacocks*. Chronicle Books, 2016. ISBN: 978-1-452-138169. Gr PreS-1

Karlins, Mark. *Music over Manhattan*. Doubleday Books for Young Readers, 1998. ISBN: 978-0-385-322256. Gr K-2

Kerley, Barbara. *A Home for Mr. Emerson*. Scholastic, 2014. ISBN: 978-054-5350-884. Gr 2-6

Lacamara, Laura. *Floating on Mama's Song*. Katherine Tegen Books, 2010. ISBN: 978-0-060-843687. Gr PreS-3

Lai, Thanhha. *Inside Out and Back Again*. HarperCollins, 2011. ISBN: 978-0-061-962783. Gr 3-7

Lewis, J. Patrick. *Tugg and Teeny*. Sleeping Bear, 2011. ISBN: 978-1-58536-514-2. Gr K-2

Lithgow, John. *Never Play Music Right Next to the Zoo*. Simon & Schuster Books for Young Readers, 2013. ISBN: 978-1-442-467439. Gr PreS-1

Look, Lenore. *Brush of the Gods*. Schwartz and Wade, 2013. ISBN: 978-0-375-87001-9. Gr K-3

MacLachlan, Patricia. *The Iridescence of Birds: A Book about Henri Matisse*. Roaring Brook, 2014. ISBN: 978-1-59643-948-1. Gr PreS-3

Markel, Michelle. *The Fantastic Jungles of Henri Rousseau*. Eerdmans, 2012. ISBN: 978-0-80285-364-6. Gr 2-6

Marsalis, Wynton. *Squeak! Rumble! Whomp! Whomp! Whomp! A Sonic Adventure*. Candlewick Press, 2012. ISBN: 978-0-7636-3991-4. Gr 1-4

McCarthy, Peter. *Jeremy Draws a Monster*. Henry Holt and Co., 2009. ISBN: 978-0-80506-934-1. Gr PreS-1

McCully, Emily Arnold. *The Secret Cave: Discovering Lascaux*. Heinemann-Raintree, 2010. ISBN: 978-0-374-36694-0. Gr 1-3

Morales, Yuyi. *Viva Frida*. Roaring Brook, 2014. ISBN: 978-1-59643-603-9. Gr PreS-3

Myers, Christopher. *Jazz*. Holiday House, 2006. ISBN: 978-0-823-415458. Gr 3-6

Newman, Leslea. *Ketzel, the Cat Who Composed*. Candlewick, 2015. IS3BN: 978-0-763-665555. Gr PreS-2

Nolan, Nina. *Mahalia Jackson: Walking with Kings and Queens*. HarperCollins, 2015. ISBN: 978-0-06087-944-0. Gr 1-2

Novesky, Amy. *Georgia in Hawaii: When Georgia O'Keefe Painted What She Pleased*. Harcourt, 2012. ISBN: 978-0-15205-420-5. Gr 2-4

Novesky, Amy. *Imogen: The Mother of Modernism and Three Boys*. Cameron + Company, 2012. ISBN: 978-1-937-359324. Gr PreS-3

Nursery Rhyme Comics: 50 Timeless Rhymes from 50 Celebrated Cartoonists. First Second, 2011. ISBN: 978-1-5964-3600-8. Gr PreS-3

Orgill, Roxane. *If I Only Had a Horn: Young Louis Armstrong*. HMH Books for Young Readers, 2002. ISBN: 978-0-618-250769. Gr 2-7

Orgill, Roxane. *Skit-Scat Raggedy Cat: Ella Fitzgerald*. Candlewick, 2012. ISBN: 978-0-763-664589. Gr 3-7

Parker, Marjorie Blain. *Colorful Dreamer: The Story of Artist Henri Matisse*. Dial Books, 2012. ISBN: 978-0-803-7375. Gr K-3

Peacock, Shane. *The Artist and Me*. Owlkids, 2016. ISBN: 978-1-771-471381. Gr K-3

Pinkney, Andrea. *Duke Ellington: The Piano Prince and His Orchestra*. Hyperion Book CH, 2006. ISBN: 978-0-7868-1420-6. Gr 1-5

Pinkney, Andrea. *Martin and Mahalia: His Words, Her Song*. Little, Brown Books for Young Readers, 2013. ISBN: 978-0-316-070133. Gr 2-4

Pinkney, Brian. *Max Found Two Sticks*. Simon and Schuster Books for Young Readers, 1994. ISBN: 978-0-671-787769. Gr K-3 Classic

Polacco, Patricia. *The Art of Miss Chew*. Putnam Juvenile, 2012. ISBN: 978-0-3992-5703-2. Gr 1-4

Powers, J. L. *Colors of the Wind: The Story of Blind Artist and Champion Runner George Mendoza*. Purple House, 2014. ISBN: 978-1-93090-073-8. Gr 2-4

Price, Leontyne. *Aida*. Gulliver Books, 1990. ISBN: 978-0-152-004057. OP

Raczka, Bob. *Action Figures: Paintings of Fun, Daring, and Adventure*. Millbrook Press, 2009. ISBN: 978-0-7613-4140-6. Gr 1-6

Raczka, Bob. *The Cosmobiography of Sun Ra: The Sound of Joy Is Enlightening*. Candlewick, 2014. ISBN: 978-0-76365-806-9. Gr 1-4

Rappaport, Doreen. *John's Secret Dreams: The John Lennon Story*. Hyperion Books, 2004. ISBN: 978-0-786-808175. Gr 1-3

Raschka, Christopher. *Charlie Parker Played Be Bop*. Scholastic Inc., 1997. ISBN: 978-0-531-070956. Gr K-3

Reich, Susanna. *Jose! Born to Dance: The Story of Jose Limon*. Simon & Schuster, 2005. ISBN: 978-0-6898-6576-3. Gr 1 and Up

Reich, Susanna. *Fab Four Friends: The Boys Who Became the Beatles*. Henry Holt and Co., 2015. ISBN: 978-0-805-094589. Gr 2-5

Reynolds, Peter H. *The Dot*. Candlewick, 2003. ISBN: 978-0-7636-1961-9. Gr PreS-4

Rosenstock, Barb. *The Noisy Paint Box: The Colors and Sounds of Kandinsky's Abstract Art*. Knopf, 2014. ISBN: 978-030-7978-493. Gr 1-3

Rusch, Elizabeth. *For the Love of Music: The Remarkable Story of Maria Anna Mozart*. Tricycle Press, 2011. ISBN: 978-1-582-463261. Gr 2-4

Ryan, Pam Munoz. *When Marian Sang*. Scholastic Press, 2002. ISBN: 978-0-4392-6967-4. Gr 1-4

Salvador, Ana. *Draw with Pablo Picasso*. Frances Lincoln Children's Books, 2008. ISBN: 978-1-8450-7819-5. Gr 1 and Up

Salzmann, Mary Elizabeth. *What in the World Is a Clarinet?* Super Sandcastle, 2012. ISBN: 978-1-61783-203-1. Gr K-2

Salzmann, Mary Elizabeth. *What in the World Is a Piano?* Super Sandcastle, 2012. ISBN: 978-1-61783-207-9. Gr K-2

Schubert, Leda. *Listen: How Pete Seeger Got America Singing*. Roaring Brook Press, 2017. ISBN: 978-1-626-722507. Gr K-3

Schubert, Leda. *Monsieur Marceau: Actor without Words*. Roaring Brook Press, 2012. ISBN: 978-1-547-18210-0. Gr PreS-3

Schwartz, Jeffrey. *The Rock and Roll Alphabet*. Mojo Hand LLC, 2011. ISBN: 978-0-6154-9521-7. Gr 1 and Up

Shapiro, J. H. *Magic Trash: A Story of Tyree Guyton and His Art*. Charlesbridge, 2015. ISBN: 978-1-580-893862. Gr K-3

Siegel, Siena Cherson. *To Dance: A Ballerina's Graphic Novel*. Atheneum Books for Young Readers, 2006. ISBN: 978-1-416-926870. Gr 4 and Up

Singer, Marilyn. *Feel the Beat: Dance Poems That Zing from Salsa to Swing*. Dial Books, 2017. ISBN: 978-0-803-740211. Gr K-3

Singer, Marilyn. *Follow, Follow*. Dutton, 2013. ISBN: 978-0-8037-3769-3. Gr 1 and Up

Singer, Marilyn. *Mirror Mirror: A Book of Reversible Verse*. Dutton, 2010. ISBN: 978-0-5254-7901-7. Gr 3-6

Singer, Marilyn. *Tallulah's Solo*. Clarion Books, 2012. ISBN: 978-0-547-330044. Gr PreS-2 Others

Smith, Cynthia Leitich. *Jingle Dancer*. HarperCollins, 2000. ISBN: 978-0-688-162412. Gr K-3

Snyder, Laura. *Swan: The Life and Dance of Anna Pavlova*. ISBN: 978-1-452-118901. Gr 1-4

Stanley, Diane. *Mozart: The Wonder Child: A Puppet Play in Three Acts*. HarperCollins, 2009. ISBN: 978-0-0607-2674-4. Gr K-3

Steptoe, Javaka. *Radiant Child: The Story of Young Artist Jean-Michel Basquiat*. Little, Brown Books for Young Readers, 2016. ISBN: 978-0-316-213882. Gr 1-5

Steptoe, John. *Mufaro's Beautiful Daughters: An African Tale*. Lothrop, Lee & Shepard, 1987. ISBN: 978-0-688-040451. Gr PreS-3

Stone, Tanya Lee. *Sandy's Circus: A Story about Alexander Calder*. Viking Books for Young Readers, 2008. ISBN: 978-0-670-062685. Gr 1-3

Stringer, Lauren. *When Stravinsky Met Nijinsky: Two Artists, Their Ballet, and One Extraordinary Riot*. Harcourt Children's Books, 2013. ISBN: 9789-0-5479-0725-3. Gr K-3

Sweet, Melissa. *Balloons over Broadway: The True Story of the Puppeteer of Macy's Parade*. Houghton, 2011. ISBN: 978-0-547-19945-0. Gr PreS-3

Tallchief, Maria. *Tallchief: America's Prima Ballerina*. Puffin Books, 2001. ISBN: 978-0-142-300183. Gr 2-4

Thompson, Bill. *Chalk*. Two Lions, 2010. ISBN: 978-0-76145-526-4. Gr PreS-4

Tonatiuh, Duncan. *Diego Rivera: His World and Ours*. Harry N. Abrams, 2011. ISBN: 978-0-810-997318. Gr 1-4

Tullet, Herve. *Press Here*. Chronicle Books, 2011. ISBN: 978-0-8118-7954-5. Gr PreS and Up

Tullet, Herve. *Mix It Up!* Chronicle Books, 2014. ISBN: 978-1-45213-735-3. Gr PreS-K

Uegaki, Chieri. *Hana Hashimoto, Sixth Violin*. Kids Can Press, 2014. ISBN: 978-1-89478-633-1. Gr K-1

Wahl, Jan. *The Art Collector*. Charlesbridge, 2011. ISBN: 978-1-58089-270-4. Gr PreS-3

Warjin, Kathy-Jo. *M Is for Melody: A Music Alphabet*. Sleeping Bear Press, 2006. ISBN: 978-1-5853-6332-2. Gr 1 and Up

Watson, Renee. *Harlem's Little Blackbird: The Story of Florence Mills*. Random House Books for Young Readers, 2012. ISBN: 978-0-375-869730. Gr PreS-2

Weatherford, Carole Boston. *Before John Was a Jazz Giant: A Song of John Coltrane*. Henry Holt and Co., 2008. ISBN: 978-0-805-079944. Gr K-4

Weatherford, Carole Boston. *Leontyne Price: Voice of a Century*. Random/Knopf, 2014. ISBN: 978-0-37585-606-8. Gr 2-5

Wiesner, David. *Art and Max*. Clarion Books, 2010. ISBN: 978-0-6187-5663-6. Gr K-4

Wiesner, David. *Flotsam*. Clarion Books, 2006. ISBN: 978-0-618-194575. Gr PreS-3

Williams-Garcia, Rita. *Bottle Cap Boys Dancing on Royal Street*. Marimba Books, 2015. ISBN: 978-1-603-490306. Gr 2-4

Winter, Jeanette. *Henri's Scissors*. Beach Lane, 2013. ISBN: 978-1-44246-484-1. Gr K-2

Winter, Jeanette. *Just Behave, Pablo Picasso!* Arthur A. Levine Books, 2012. ISBN: 978-0-545-13291-6. Gr 1-4

Winter, Jeanette. *Mr. Cornell's Dream Boxes*. Beach Lane, 2014. ISBN: 978-144-2499-003. Gr K-2

Wood, Susan. *Esquivel! Space-Age Sound Artist*. Charlesbridge, 2016. ISBN: 978-1-580 -8966733. Gr 2-6

Young, Ed. *The House Baba Built: An Artist's Childhood in China*. Little, Brown Books for Young Readers, 2011. ISBN: 978-0-3160-7628-9. Gr 1 and Up

Informational

Abel, Jessica. *Drawing Words & Writing Pictures: Making Comics: Manga, Graphic Novels and Beyond*. First Second, 2008. ISBN: 978-1-5964-3131-7. YA

Adkins, Jan. *Up Close: Frank Lloyd Wright*. Viking, 2007. ISBN: 978-0-670-06138-9. YA

Amara, Philip. *So You Want to Be a Comic Book Artist?* Simon & Schuster Children's Publishing, 2012. ISBN: 978-1-58270-358-9. Gr 4-6

Angleberger, Tom. *Art2-D2's Guide to Folding and Doodling: An Origami Yoda Activity Book*. Amulet Books, 2013. ISBN: 978-1-4197-0534-2. Gr 4 and Up

Arndt, Ingo. *Animal Architecture*. Harry N. Abrams, 2014. ISBN: 978-1-419-711657. Reference

Aronson, Marc. *Eyes of the World: Robert Capa, Gerda Taro, and the Invention of Modern Photojournalism*. Henry Holt and Co., 2017. ISBN: 978-0-805-098358. Reference

Art Book for Children: Book Two. Phaidon, 2007. ISBN: 978-0-7148-4706-1. Gr 4-6

Basher, Simon. *Basher Basics: Music*. Kingfisher, 2011. ISBN: 978-0-75346-595-0. Gr 3-7

Bauer, Helen. *Beethoven for Kids: His Life and Music with 21 Activities*. Chicago Review Press, 2011. ISBN: 978-1-5697-4500-7. Gr 4 and Up

Bauer, Helen. *Verdi for Kids: His Life and Music with 21 Activities*. Chicago Review Press, 2013. ISBN: 978-1-6137-4500-7. Gr 5-8

Bertholf, Bert. *The Long Gone Lonesome History of Country Music*. Little, Brown Books for Young Readers, 2007. ISBN: 978-0-316-523936. Gr 4-6

Bernier-Grand, Carmen. *Alicia Alonzo: Prima Ballerina*. Two Lions, 2011. ISBN: 978-0-761-455622. Gr 4 and Up

Bidner, Jenni. *The Kids' Guide to Digital Photography: How to Shoot, Save, Play with and Print Your Digital Photos*. Lark Books, 2004. ISBN: 978-1-579-906044. Gr 4 and Up

Bliss, John. *Art That Moves: Animation around the World*. Raintree, 2011. ISBN: 978-1-4109-3922-7. Gr 5-8

Children's Book of Art. DK Publishing, 2009. ISBN: 978-0-7566-5511-2. Gr 3 and Up

Children's Book of Music. DK Publishing, 2010. ISBN: 978-0-7566-6734-4. Gr 3-6

Chin-Lee, Cynthia. *Akira to Zoltan: Twenty-Six Men Who Changed the World*. Charlesbridge, 2008. ISBN: 978-1-57091-580-2. Gr 3-7

Christensen, Bonnie. *Fabulous! A Portrait of Andy Warhol*. Henry Holt and Company, 2011. ISBN: 978-0-8050-8753-6. Gr 1-4

Christensen, Bonnie. *Woody Guthrie: Poet of the People*. Dragonfly Books, 2009. ISBN: 978-0-5531-8753-6. Gr 1-4

Close, Chuck. *Chuck Close: Face Book*. Abrams, 2012. ISBN: 978-1-4197-0163-4. Gr 4 and Up

Cohn, Jessica. *Animator*. Gareth Stevens, 2010. ISBN: 978-1-4339-1953-4. Gr 4 and Up

Crossingham, John. *Learn to Speak Music: A Guide to Creating, Performing, and Promoting Your Songs*. Owlkids, 2009. ISBN: 978-1-8973-4965-6. Gr 5-8

Cruz, Barbara C. *Alvin Ailey: Celebrating African American Culture in Dance*. Enslow Publishers, 2004. ISBN: 978-0-766-022935. Gr 5-9

DeCarufel, Laura. *Learn to Speak Fashion: A Guide to Creating, Showcasing & Promoting Your Style*. Owlkids Books, 2012. ISBN: 978-1-9269-7337-1. Gr 6-10

DePrince, Michaela and Elaine DePrince. *Taking Flight: From War Orphan to Star Ballerina*. Knopf, 2014. ISBN: 978-0-38575-511-5. Gr 6 and Up

Devorkin, David. *The Hubble Cosmos: 25 Years of New Vistas in Space*. National Geographic, 2015. ISBN: 978-1-426-215575

Ehlert, Lois. *The Scraps Book: Notes from a Colorful Life*. Beach Lane, 2014. ISBN: 978-144-2435-711. Gr K-5

Finger, Brad. *13 American Artists Children Should Know.* Prestel, 2010. ISBN: 978-3-7913 -7036-1. Gr 2 and Up

Finger, Brad. *13 Modern Artists Children Should Know.* Prestel, 2010. ISBN: 978-3-7913 -7015-6. Gr 4 and Up

Fleischman, Sid. *Sir Charlie Chaplin: The Funniest Man in the World.* Greenwillow Books, 2010. ISBN: 978-0-06-189640-8. Gr 6-12

Geis, Alexander. *Alexander Calder: Meet the Artist!* Princeton Architectural Press, 2014. Gr 2-7

Geis, Alexander. *Henri Matisse: Meet the Artist!* Princeton Architectural Press, 2014. ISBN: 978-1-616-892821 Gr 2-7

Geis, Alexander. *Vincent van Gogh: Meet the Artist!* Princeton Architectural Press, 2015. ISBN: 978-1-616-894566. Gr 2-7

Gherman, Beverly. *Sparky: The Life and Art of Charles Schulz.* Chronicle Books, 2009. ISBN: 978-0-8118-6790-0. Gr 4-6

Golio, Gary. *Jimi Sounds Like a Rainbow: A Story of the Young Jimi Hendrix.* Clarion Books, 2010. ISBN: 978-0-6188-5279-6. Gr 6-9

Golio, Gary. *Spirit Seeker: John Coltrane's Musical Journey.* Clarion Books, 2012. ISBN: 978-0-5472-3994-1. Gr 5-8

Greenberg, Jan and Sandra Jordan. *The Mad Potter: George E. Ohr, Eccentric Genius.* Roaring Brook, 2013. ISBN: 978-1-59643-810-1. Gr 5-8

Haeckel, Ernst. *Art Forms in Nature: The Prints of Ernst Haeckel.* Prestel, 2008. ISBN; 978-3-791-319902. Reference

Heine, Florian. *13 Art Inventions Children Should Know.* Prestel, 2011. ISBN: 978-3-7913 -7060-6. Gr 4 and Up

Holmes, Marc Taro. *Designing Creatures and Characters: How to Build an Artist's Portfolio for Video Games, Film, Animation and More.* North Light Books, 2016. ISBN: 978-1-440 -344091. Adult

Hosack, Karen. *Buildings.* Raintree, 2008. ISBN: 978-1-4109-3165-8. Gr 4-8

Hosack, Karen. *Drawings and Cartoons.* Raintree, 2008. ISBN: 978-1-4109-3163-4. Gr 4-8

Jocelyn, Marthe. *Sneaky Art: Crafty Surprises to Hide in Plain Sight.* Candlewick Press, 2013. ISBN: 978-0-7636-5648-5. Gr 3-7

Kallen, Stuart. *Manga.* Lucent, 2011. ISBN: 978-1-4205-0535-1. YA

Kidd, Chip. *Go: A Kidd's Guide to Graphic Design.* Workman, 2013. ISBN: 978-0-76117-219-2. Gr 8-12

Krull, Kathleen. *Lives of the Musicians: Good Times, Bad Times (and What the Neighbors Thought).* HMH Books for Young Readers, 2011. ISBN: 978-0-15216-436-2. Gr 4-7

Krull, Kathleen. *The Beatles Were Fab (and They Were Funny).* Harcourt Children's Books, 2013. ISBN: 978-0-5475-0991-4. Gr 6-9

Krull, Kathleen. *Lives of the Artists: Masterpieces, Messes (and What the Neighbors Thought).* HMH Books for Young Readers, 2014. ISBN: 978-0-544-252233. Gr 4-7

Laroche, Giles. *What's Inside? Fascinating Structures around the World.* Houghton Mifflin, 2009. ISBN: 978-0-618-86247-4. Gr 4-8

Levete, Sarah. *Maker Projects for Kids Who Love Animation.* Crabtree, 2016. ISBN: 978-0-778-722564. Gr 3-7

Liew, Sonny. *The Art of Charlie Chan Hock Chye.* Pantheon, 2016. ISBN: 978-1-101-870693. Adult

Loewen, Nancy. *Action! Writing Your Own Play.* Picture Window Press, 2011. ISBN: 978-1-404-86392-7. Gr 2-4

Loewen, Nancy. *Show Me a Story: Writing Your Own Picture Book.* Picture Window Books, 2009. ISBN: 978-1-404-853423. Gr 2-4

Mack, Jim. *Hip-Hop.* Heinemann-Raintree, 2010. ISBN: 978-1-410-933935. Gr 3-6

Marsalis, Wynton. *Jazz ABC: An A to Z Collection of Jazz Portraits.* Candlewick, 2007. ISBN: 978-0-763 6-3434-6. Gr 7 and Up

McCloud, Scott. *Making Comics: Storytelling Secrets of Comics, Manga, and Graphic Novels.* HarperPerennial, 2006. ISBN: 978-0-06-078094-4. Gr 6 and Up

McKendry, Joe. *One Times Square: A Century of Change at the Crossroads of the World.* David R. Godine, 2012. ISBN: 978-1-5679-2364-3. Gr 4-7

McMullan, James. *Leaving China: An Artist Paints His World War II Childhood.* Algonquin Young Readers, 2016. ISBN: 978-1-616-202552. Gr 7 and Up

Miles, Liz. *Making a Recording.* Raintree, 2009. ISBN: 978-1-41093-392-8. Gr 3-6

Miles, Liz. *Movie Special Effects.* Heinemann-Raintree, 2010. ISBN: 978-1-4109-3399-7. Gr 3-6

Miles, Liz. *The Orchestra*. Heinemann-Raintree, 2010. ISBN: 978-1-4109-3394-2. Gr 3-6

Miles, Liz. *Photography*. Heinemann-Raintree, 2010. ISBN: 978-41093-417-8. Gr 3 and Up

Murphy, Claire Rudolf. *My Country, 'Tis of Thee: How One Song Reveals the History of Civil Rights*. Holt, 2014. ISBN: 978-080-5082-265. Gr 5-7

My Art Book: Amazing Art Projects Inspired by Masterpieces. DK, 2011. ISBN: 978-0-7566 -7582-0. Gr 3-6

Myers, Walter Dean. *Blues Journey*. Holiday House, 2003. ISBN: 978-0-823-416134. Gr 5-8

Neri, G. *Hello, I'm Johnny Cash*. Candlewick, 2014. ISBN: 978-0-76366-245-5. Gr 4-7

Neuburger, Emily. *Show Me a Story: 40 Craft Projects and Activities to Spark Children's Storytelling*. Storey Publishing, 2012. ISBN: 978-60342-988-7. Gr K-7

Nobleman, Marc Tyler. *Bill the Boy Wonder: The Secret Co-Creator of Batman*. Charlesbridge, 2012. ISBN: 978-1-58089-289-6. Gr 2-6

Nobleman, Marc Tyler. *Boys of Steel: The Creators of Superman*. Knopf Books for Young Readers, 2008. ISBN: 978-0-3758-3802-6. Gr 5 and Up

O'Brien, Lisa. *Lights, Camera, Action! Making Movies and TV from The Inside Out*. Maple Tree Press, 2007. ISBN: 1-897066-89-9. Gr 4-8

O'Connor, Jim. *Who Is Bob Dylan?* Grosset & Dunlap, 2013. ISBN: 978-0-4484-6461-9. Gr 4 and Up

Orgil, Roxane. *Jazz Day: The Making of a Famous Photograph*. Candlewick, 2016. ISBN: 978-0-763-669546. Gr 3-7

Phelan, Matt. *Bluffton*. Candlewick, 2013. ISBN: 978-076-3650-797. Gr 3-6

Powell, Patricia Hruby. *Josephine: The Dazzling Life of Josephine Baker*. Chronicle, 2014. ISBN: 978-1-45210-314-3. Gr 2-5

Raczka, Bob. *Before They Were Famous: How Seven Artists Got Their Start*. Millbrook, 2010. ISBN: 978-0-76136-077-3. Gr 2-6

Robertson, Robbie. *Legends, Icons and Rebels: Music That Changed the World*. Tundra Books, 2013. ISBN: 978-1-770-495715 Gr 4-7

Roche, Art. *Art for Kids: Cartooning: The Only Cartooning Book You'll Ever Need to Be the Artist You've Always Wanted to Be*. Sterling, 2010. ISBN: 978-1-4020775154. Gr 3-12

Roche, Art. *Art for Kids: Comic Strips: Create Your Own Comic Strips from Start to Finish.* Lark Books, 2007. ISBN: 978-1-5799-0788-4. Gr 4 and Up

Roeder, Annette. *13 Buildings Children Should Know.* Prestel, 2009. ISBN: 978-3-7913-4171-2. Gr 5 and Up

Rubalcaba, Jill. *I. M. Pei: Architect of Time, Space, and Purpose.* Marshall Cavendish, 2011. ISBN: 978-0-7614-5973-6. YA

Rubin, Susan Goldman. *Diego Rivera: An Artist for the People.* Abrams, 2013. ISBN: 978-0-81098-411-0. Gr 5-8

Rubin, Susan Goldman. *Everybody Paints! The Lives and Art of the Wyeth Family.* Chronicle Books, 2014. ISBN: 978-0-811-869843. Gr 7 and Up

Rubin, Susan Goldman. *Music Was It: Young Leonard Bernstein.* Charlesbridge Pub Inc., 2011. ISBN: 978-1-5808-9344-2. Gr 5 and Up

Rubin, Susan Goldman. *Music Was It: Young Leonard Bernstein.* Charlesbridge, 2015. ISBN: 978-1-580-893459. Gr 6-9

Rubin, Susan Goldman. *Whaam! The Art and Life of Roy Lichtenstein.* Abrams, 2009. ISBN: 978-0-81099-492-8. Gr 2-6

Russell-Brown, Katheryn. *Little Melba and Her Big Trombone.* Lee & Low, 2014. ISBN: 978-1-60060-898-8. Gr 1-5

Sabbeth, Carol. *Van Gogh and the Post-Impressionists for Kids: Their Lives and Ideas.* Chicago Review Press, 2011. ISBN: 978-1-5697-6275-2. Gr 5 and Up

Say, Allen. *Drawing from Memory.* Scholastic Press, 2011. ISBN: 978-0-5451-7686-6. Gr 5 and Up

Shaskan, Trisha Speed. *Art Panels, BAM! Speech Bubbles, POW! Writing Your Own Graphic Novel.* Picture Window Books, 2011. ISBN: 978-1-4048-6393-4. Gr 2-4

Slade, Suzanne. *The Music in George's Head, George Gershwin Creates Rhapsody in Blue.* Calkins Creek, 2016. ISBN: 978-1-629-790992. Gr 3-6

Stamaty, Mark Alan. *Shake, Rattle and Turn That Noise Down: How Elvis Shook Up Music, Me and Mom.* Knopf Books for Young readers, 2010. ISBN: 978-0-375-846854. Gr 4-8

Stoneham, Bill. *How to Create Fantasy Art for Video Games: Complete Guide to Creating Concepts, Characters, and Worlds.* Barron's, 2010. ISBN: 978-0-7641-4504-9. Adult

Strum, James. *Adventures in Cartooning: How to Turn Your Doodles into Comics.* Roaring Brook, 2009. ISBN: 978-1-5964-3369-4. Gr 6-8

Tan, Shaun. *The Bird King: An Artist's Notebook.* Arthur A. Levine, 2013. ISBN: 978-0-5454-6513-7. Gr 3 and Up

Tate, Don. *It Jes' Happened: When Bill Traylor Started to Draw.* Lee & Low, 2012. ISBN: 978-1-6006-0260-3. Gr. 2-4

Underwood, Deborah. *Staging a Play.* Raintree, 2009. ISBN: 978-1-4109-3396-6. Gr 3-6

Van Hecke, Susan. *Raggin' Jazzin' Jazzin': A History of American Musical Instrument Makers.* Boyds Mills Press, 2011. ISBN: 978-1-59078-574-4. Gr 6 and Up

Watt, Fiona. *Step-by-Step Drawing Book.* Usborne Books, 2014. ISBN: 978-0-794-529536. Gr 4 and Up

Wenzel, Angela. *13 Art Mysteries Children Should Know.* Prestel, 2011. ISBN: 978-3-7913-7044-6. Gr 5-7

Wenzel, Angela. *13 Artists Children Should Know.* Prestel, 2009. ISBN: 978-3-7913-4173-6. Gr 4 and Up

Wenzel, Angela. *13 Sculptures Children Should Know.* Prestel, 2010. ISBN: 978-3-7913-7010-1. Gr 4 and Up

Williams, Freddie. E. *The DC Comics Guide to Digitally Drawing Comics.* Watson-Guptill, 2009. ISBN: 978-0-8230-9923-8. Gr 9-12

Williams-Garcia, Rita. *Clayton Byrd Goes Underground.* Amistad, 2017. ISBN: 978-0-062-215918. Gr 3-7

Wing, Natasha. *An Eye for Color: The Story of Josef Albers.* Holt, 2009. ISBN: 978-0-80508-072-8. Gr 2-6

Wray, Anna. *This Belongs to Me: Cool Ways to Personalize Your Stuff.* RP Kids, 2013. ISBN: 978-0-76244-929-3. Gr 3 and Up

Series

A Kids Guide to Drawing (Rosen) Several Sets of Series

All About Colors (Enslow)

Artists Through the Ages (Rosen)

Awesome Art (Rosen)

Children Should Know (Prestel)

Clever Crafts (Rosen)

Crafts and Cultures of the Middle Ages (Rosen)

Culture in Action (Raintree)

Do It Yourself Projects (Rosen)

Drawing Series (Rosen) Several Sets

How to Draw Series (Rosen) Several Sets

It's Fun to Draw Series (Rosen) Several Sets

Let's Draw Series (Rosen) Several Sets

Meet the Artist (Rosen)

Musical Instruments (Super Sandcastle)

Stories in Art (Rosen)

Who Was . . .? (Grosset & Dunlap)

Young Artists of the World (Rosen)

Chapter 10

MATH

Mathematics is the music of reason.

—James Joseph Sylvester
http://www.brainyquote.com

Mathematics *includes algebra, geometry, measurement, data analysis, probability, problem-solving, reasoning, proofs, communication theory, calculus, trigonometry, symbolic relationships, patterns, shapes and reasoning. Math is the science of patterns and relationships; it provides an exact language for technologies, science, and engineering.*
—Dugger, William E. "STEM: Some Basic Definitions"
(Senior Fellow, International Technology and
Engineering Educators Association).
http://www.iteea.org

GENERAL QUESTIONS

What did you learn from this book?

What qualifications did the author have to write this book?

Where can you go to find out more information about this topic?

What was the purpose of this book?

What does the author want the reader to believe about this topic?

Was there anything in this book that you did not understand?

FEATURED AUTHORS AND ANNOTATIONS

Name: Cindy Neuschwander

Place of Birth: San Diego, California

About: Cindy Neuschwander's father was a naval officer and later, a high school teacher, and her mother was a homemaker. She has one younger brother. Cindy graduated with a BA in International Studies from Willamette University and earned an MA from Stanford University. She has taught all grades in elementary school as well as high school. In addition to her teaching, Cindy is the author of eight published picture books for children with mathematical themes. Cindy began writing books in 1994. She had used math literature with her own classes in the early 1990s and liked the way students responded to it. She wanted to use more of these books but found there were not many available, so she started writing some of her own.

Website: https://www.charlesbridge.com/pages/cindy-neuschwander

Annotated Title:

Nueschwander, Cindy. *Sir Cumference and the Viking's Map*. Charlesbridge, 2012. ISBN: 978-1-57091-791-2.

> Per and her cousin, Radius, find a Viking map and must figure out what the X and Y axes mean on the number grid. They have many adventures along the way as they search for the treasure.

Name: Greg Tang

Place of Birth: Ithaca, New York

About: Greg Tang is the author of eight picture books that teach math. He is a middle/high school math teacher. His books have been on the *New York Times* Children's Books Best Seller List and have won many awards. He provides workshops for teachers and author visits for schools.

Website: http://gregtangmath.com

Annotated Title:

Tang, Greg. *The Grapes of Math*. Scholastic, 2001. ISBN: 978-0-43959-840-8. Gr 2-5

> This was Greg Tang's first picture book about math. He wrote it to help kids see numbers as groups rather than counting by ones all the time. Rhymes and riddles abound in this book—to help kids to group objects together rather than counting one by one. This is a great introduction to problem-solving.

Name: Sarah C. Campbell

Place of Birth: Evanston, Illinois

About: Before she began writing books, Sarah covered politics, business, government, and courts for two daily newspapers. She taught journalism and wrote annual reports and other public relations pieces. She created a grant-funded program at her kids' elementary school to bring artists into the classroom.

Website: http://www.sarahccampbell.com/

Annotated Title:

Campbell, Sarah C. *Growing Patterns: Fibonacci Numbers in Nature*. Boyds Mills Press, 2010. Gr K-3

The Fibonacci numbers sequence (1, 1, 2, 3, 5, 8, 13) is found in nature: a nautilus shell, a pineapple, a sunflower, and a pine cone—for example. This book provides some amazingly beautiful photographs of the spirals in nature.

ACTIVITIES THAT CONNECT MATH

Fibonacci

Science	Technology	Engineering
Fibonacci sequence in nature—http://io9.com/ 5985588/15-uncanny-examples-of-the-golden-ratio-in-nature	3-D Printed spiral jewelry —http://www.instructables .com/id/3D-Printed-Spiral-Earrings/	Make a spiral wind spinner —https://www.youtube.com /watch?v=3s1RPpz8P7I

Arts	Math
How to draw the golden spiral— http://thehelpfulartteacher .blogspot.com/2012/01/spiral.html	Campbell, Sarah C. *Growing Patterns: Fibonacci Numbers in Nature.* Boyds, 2010. Gr K-3

INVENTIONS AND INVENTORS

Abdul-Jabbar, Kareem. *What Color Is My World? The Lost History of African-American Inventors.* Candlewick, 2013. ISBN: 978-0-76366-442-8. Gr 3-7

Ball, Jacqueline A. *Communication Inventions: From Hieroglyphics to DVDs.* Bearport Publishing, 2005. ISBN: 978-1-59716-129-9. Gr 2 and Up

Barton, Chris. *The Day-Glo Brothers: The True Story of Bob and Joe Switzer's Bright Ideas and Brand-New Colors.* Charlesbridge, 2009. ISBN: 978-1-57091-673-1. Gr 2-5

Murphy, Glenn. *Inventions.* Simon & Schuster Books for Young Readers, 2009. ISBN: 978-1-416-93865-1. Gr 3-7

Rhatigan, Joe. *Inventions That Could Have Changed the World . . . But Didn't.* Imagine, 2015. ISBN: 978-1-62354-024-1. Gr 3-7

Thimmesh, Catherine. *Girls Think of Everything: Stories of Ingenious Inventions by Women.* HMH Books for Young Readers, 2002. ISBN: 978-0-61819-563-3. Gr 4-6

BIBLIOGRAPHY

Picture Books

Adler, David A. *Money Madness.* Holiday House, 2009. ISBN: 978-0-823-414741. Gr K-3

Adler, David A. *Triangles.* Holiday House, 2015. ISBN: 978-0-823-433056. Gr 2-5

Anderson, Jill. *Finding Shapes with Sebastian Pig and Friends at the Museum*. Enslow, 2009. ISBN: 978-0-7660-3363-4. Gr 1-3

Anderson, Jill. *Money Math with Sebastian Pig and Friends at the Farmer's Market*. Enslow, 2009. ISBN: 978-0-7660-3364-1. Gr PreS and Up

Anno, Mitsumasa. *Anno's Mysterious Multiplying Jar*. Penguin Putnam Books for Young Readers, 1999. ISBN: 978-0-69811-753-2. Gr PreS-3 Classic

Banks, Kate. *Max's Math*. Farrar, Straus and Giroux, 2015. ISBN: 978-0-374-348755. Gr PreS-3

Barnett, Mac. *Count the Monkeys*. Scholastic, 2013. ISBN: 978-0-545-641487. Gr PreS-1 OP

Berkes, Marianne. *Over in the Forest: Come and Take a Peek*. Dawn Publications, 2012. ISBN: 978-1-584-691631. Gr PreS-1. Others

Brown, Anthony. *One Gorilla: A Counting Book*. Candlewick, 2015. ISBN: 978-0-763-679156. Gr PreS and Up

Bullard, Lisa. *Brody Borrows Money*. Millbrook Press, 2013. ISBN: 978-1-467-715089. Gr K-2. Others

Burns, Marilyn. *The Greedy Triangle*. Scholastic Press, 1995. ISBN: 978-0-590-489911. Gr PreS-3

Calvert, Pam. *Multiplying Menace: The Revenge of Rumpelstiltskin*. Charlesbridge, 2006. ISBN: 1-57091-889-9. Gr K-3

Campbell, Sarah C. *Mysterious Patterns: Finding Fractals in Nature*. Boyds Mills Press, 2014. ISBN: 978-1-620-916278. Gr 2-5

Cave, Kathryn. *Out for the Count: A Counting Adventure*. Simon & Schuster Children's Publishing, 2006. ISBN: 978-0-6717-5591-1. Gr PreS-2

Cleary, Brian P. *A Fraction's Goal—Parts of a Whole*. Millbrook, 2011. ISBN: 978-0-8225-7881-9. Gr K-3

Cleary, Brian P. *A Dollar, a Penny, How Much and How Many?* Millbrook Press, 2012. ISBN: 978-0-8225-7882-6. Gr K-3

Clements, Andrew. *A Million Dots*. Simon & Schuster, 2006. ISBN: 978-0-6898-5824-6. Gr 1-4

Cotton, Katie. *Counting Lions: Portraits from the Wild*. Candlewick, 2015. ISBN: 978-0-763-682071. Gr K-12

Cuyler, Margery. *Guinea Pigs Add Up*. Walker, 2010. ISBN: 978-0-8027-9795-7. Gr PreS-2

Daniels, Teri. *Math Man*. Orchard Books, 2001. ISBN: 978-0-43929-308-2. Gr K-3

Demi. *One Grain of Rice*. Scholastic, 1997. ISBN: 978-0-59093-998-0. Gr PreS-3. Classic

Dodds, Dayle Ann. *Minnie's Diner: A Multiplying Menu*. Candlewick, 2004. ISBN: 0-7636-1736-9. Gr K-3

Ehlert, Lois. *Fish Eyes: A Book You Can Count On*. HMH Books for Young Readers, 1990. ISBN: 978-0-15228-050-5. Gr PreS-1 Classic

Franco, Betsy. *Mathematickles!* Simon and Schuster Children's Publishing, 2003. ISBN: 0-689-84357-7. Gr K-5

Franco, Betsy. *Zero Is the Leaves on the Tree*. Tricycle, 2009. ISBN: 978-1-58246-249-3. Gr K-3

Fromental, Jean-Luc. *365 Penguins*. Abrams Books for Young Readers, 2006. ISBN: 978-0-8109-4460-2. Gr K-4

Goldstone, Bruce. *I See a Pattern Here*. Henry Holt and Co., 2015. ISBN: 978-0-805-092097. Gr 2-4

Goldstone, Bruce. *100 Ways to Celebrate 100 Days*. Henry Holt, 2010. ISBN: 978-0-8050-8987-4. Gr PreS-2

Gravett, Emily. *The Rabbit Problem*. Simon and Schuster Children's Publishing, 2010. ISBN: 978-14424-1255-2. Gr 1-5

Haskins, James. *Count Your Way through Afghanistan*. Millbrook, 2006. ISBN: 978-1-575-058801. Gr 2 and Up

Heiligman, Deborah. *The Boy Who Loved Math: The Improbable Life of Paul Erdos*. Roaring Brook Press, 2013. ISBN: 978-1-59643-307-6. Gr PreS-2

Holub, Joan. *Zero the Hero*. Henry Holt and Co., 2012. ISBN: 978-0-8050-9384-1. Gr 1-5

Hopkins, Lee Bennett. *Marvelous Math: A Book of Poems*. Simon & Schuster, 2001. ISBN: 0-689-80658-2. Gr K-3

Hosford, Kate. *Infinity and Me*. Carolrhoda, 2012. ISBN: 978-0-8225-7882-6. Gr K-4

Hutchins, Hazel. *A Second Is a Hiccup*. Arthur A. Levin Books, 2004. ISBN: 978-0-4398-3106-2. Gr K-1

Jackson, Ellen. *Octopuses One to Ten*. Beach Lane Books, 2016. ISBN: 978-1-481-431828. Gr PreS-3

Jenkins, Emily. *Five Creatures*. Farrar, 2001. ISBN: 0-374-32341-0. Gr K-3

Jenkins, Emily. *Lemonade in Winter: A Book about Two Kids Counting Money*. Random, 2012. ISBN: 978-0-3758-5883-2. Gr PreS-2

Kang, Anna. *You Are (Not) Small*. Two Lions, 2014. ISBN: 978-1-477-847725. Gr PreS-1

Larochelle, David. *1+1=5: And Other Unlikely Additions*. Sterling, 2010. ISBN: 978-1-4027-5995-6. Gr K-3

Lasky, Kathryn. *Librarian Who Measured the Earth*. Little, Brown Books, 1994. ISBN: 978-0-316-515269. Gr PreS-1

Lawler, Janet. *Ocean Counting*. National Geographic Children's Books, 2013. ISBN: 978-1-426-31116-1. Gr PreS-1

Leedy, Loreen. *The Great Graph Contest*. Holiday House, 2006. ISBN: 978-0-823-420292. Gr K-3

Leedy, Loreen. *It's Probably Penny*. Henry Holt, 2007. ISBN: 978-0-8050-7389-8. Gr K-3

Leedy, Loreen. *Measuring Penny*. National Geographic School Pub, 2010. ISBN: 978-0-805-065725. Gr K-3

Leedy, Loreen. *Missing Math: A Number Mystery*. Two Lions, 2015. ISBN: 978-1-477-810927. Gr K-3

Leedy, Loreen. *Seeing Symmetry*. Holiday House, 2012. ISBN: 978-0-8234-2360-6. Gr 2-4.

Lewis, J. Patrick. *Arithme-Tickle: An Even Number of Odd Riddle-Rhymes*. Harcourt Children's Books, 2002. ISBN: 978-0-152-164188. Gr PreS-3

Lewis, J. Patrick. *Edgar Allan Poe's Pie: Math Puzzlers in Classic Poems*. HMH Books for Young Readers, 2015. ISBN: 978-0-544-456129. Gr 1-4

Lionni, Leo. *Inch by Inch*. HarperTrophy, 1995. ISBN: 978-0-6881-3283-5. Gr K-3 Classic

Litwin, Eric. *Pete the Cat and His Four Groovy Buttons*. HarperCollins, 2012. ISBN: 978-0-062-110589. Gr PreS-2

Long, Ethan. *The Wing Wing Brothers Math Spectacular*. Holiday House, 2012. ISBN: 978-0-8234-2320-0. Gr K-1

Maloney, Peter. *One Foot Two Feet: An EXCEPTIONal Counting Book*. Putnam, 2011. ISBN: 978-0-399-25446-8. Gr PreS-K

Mariconda, Barbara. *Ten for Me*. Sylvan Dell Publishing, 2011. ISBN: 978-1-607-180746. Gr K-3

Markel, Michelle. *Tyrannosaurus Math*. Tricycle, 2009. ISBN: 978-1-58246-282-0. Gr PreS-2

Marzollo, Jean. *Help Me Learn Numbers 0–20*. Holiday House, 2011. ISBN: 978-0-8234-2334-7. Gr PreS-1

McNamara, Margaret. *How Many Seeds in a Pumpkin?* Schwartz & Wade, 2007. ISBN: 978-0-37584-014-2. Gr PreS-2

Medearis, Angela Shelf. *The 100th Day of School*. Cartwheel, 1996. ISBN: 978-0-590-259446. Gr PreS-3

Medina, Juana. *1 Big Salad: A Delicious Counting Book*. Viking Books for Young Readers, 2016. ISBN: 978-1-101-999745. Gr PreS-1

Menotti, Andrea. *How Many Jelly Beans?* Chronicle, 2012. ISBN: 978-1-4521-0206-1. Gr K-4

Merriam, Eve. *Twelve Ways to Get to Eleven*. Simon & Schuster Children's Publishing, 1996. ISBN: 0-671-75544-7. Gr PreS and Up

Murphy, Stuart J. *Captain Invincible and the Space Shapes*. HarperCollins, 2001. ISBN: 978-0-064-467315. Gr K-4

Murphy, Stuart J. *Tally O'Malley*. HarperCollins, 2004. ISBN: 978-0-060-531645. Gr 1-5

Murray, Diana. *City Shapes*. Little, Brown Books for Young Readers, 2016. ISBN: 978-0-316-370929. Gr PreS-3

Nelson, Robin. *Let's Make a Bar Graph*. Lerner Publications, 2013. ISBN: 978-0-7613-8972-9. Gr K-3

Nelson, Robin. *Let's Make a Tally Chart*. Lerner Publications, 2013. ISBN: 978-0-7613-8975-0. Gr K-3

Neuschwander, Cindy. *Sir Cumference and the Dragon of Pi: A Math Adventure*. Charlesbridge, 1999. ISBN: 978-1-57091-164-4. Gr 2-7 Classic

Neuschwander, Cindy. *Sir Cumference and the Great Knight of Angleland: A Math Adventure*. Charlesbridge, 2001. ISBN: 978-1-57091-169-9. Gr 1-7 Classic

Newman, Leslea. *Ketzel, the Cat Who Composed*. Candlewick, 2015. ISBN: 978-0-763-665555. Gr PreS-2

Otoshi, Kathryn. *One*. KO Kids Books, 2008. ISBN: 978-0-972-394642. Gr PreS-K

Otoshi, Kathryn. *Zero*. KO Kids Books, 2010. ISBN: 978-0-972-394635. Gr 1-2 Others

Overdeck, Laura. *Bedtime Math: A Fun Excuse to Stay Up Late*. Feiwel and Friends, 2013. ISBN: 978-1-250-03595-1. Gr PreS-2

Overdeck, Laura. *Bedtime Math 2: This Time It's Personal*. Feiwel and Friends, 2014. ISBN: 978-1-250-040961. Gr PreS-4 Others

Robbins, Ken. *For Good Measure: The Ways We Say How Much, How Far, How Big*. Roaring Brook, 2010. ISBN: 978-1-59643-344-1. Gr PreS-3

Robinson, Fiona. *Ada's Ideas: The Story of Ada Lovelace, the World's First Computer Programmer*. Abrams Books for Young Readers, 2016. ISBN: 978-1-419-718724. Gr 1-4

Robinson, Tom. *Fibonacci Zoo*. Arbordale Publishing, 2015. ISBN: 978-1-628-555662. Gr 1-4

Rosenthal, Amy Krouse. *Wumbers*. Chronicle, 2012. ISBN: 978-1-4521-1022-6. Gr K-4

Schaefer, Lola M. *Lifetime: The Amazing Numbers in Animal Lives*. Chronicle Books, 2013. ISBN: 978-1-452-10714-9. Gr PreS-3

Schwartz, David M. *How Much Is a Million?* HarperCollins, 2004. ISBN: 978-0-68809-933-6. Gr K-3 Classic

Schwartz, David M. *Millions to Measure*. HarperCollins, 2003. ISBN: 0-688-12916-1. Gr K-3

Scieszka, Jon. *Math Curse*. Viking, 1995. ISBN: 978-0-67086-194-1. Gr 2 and Up Classic

Sebe, Masayuki. *Let's Count to 100!* Kids Can, 2011. ISBN: 978-1-55453-661-0. Gr PreS-2

Shulevitz, Uri. *One Monday Morning*. Farrar, 2003. ISBN: 978-0-3744-5648-1. Gr K-1

Singleton, Linda Joy. *Cash Kat*. Arbordale Publishing, 2016. ISBN: 978-1-628-557282. Gr K-3

Slade, Suzanne. *The Great Divide*. Sylvan Dell Publishing, 2012. ISBN: 978-1-60718-548-2. Gr 1 and Up

Slade, Suzanne. *Multiply on the Fly*. Sylvan Dell Publishing, 2011. ISBN: 978-1-60718-128-6. Gr 1 and Up

Smith, David. J. *If: A Mind-Bending New Way of Looking at Big Ideas and Numbers*. Kids Can Press, 2014. ISBN: 978-1-894-786348. Gr 2-5

Smith, Marie. *T Is for Time*. Sleeping Bear Press, 2015. ISBN: 978-1-585-365128. Gr 2-5

Stanley, Diane. *Ada Lovelace, Poet of Science*. Simon & Schuster, 2016. ISBN: 978-1-481-452496. Gr K-3

Wallmark, Laurie. *Ada Byron Lovelace and the Thinking Machine*. Creston Books, 2015. ISBN: 978-1-939-547200. Gr 1-4

Wells, Rosemary. *Bunny Money*. Puffin Books, 2000. ISBN: 978-0-140-567502. Gr PK-2

Wells, Rosemary. *Emily's First 100 Days of School*. Disney-Hyperion, 2005. ISBN: 978-0-786-813544. Gr PreS-1

Yolen, Jane. *Count Me a Rhyme: Animal Poems by Numbers*. WordSong, 2014. ISBN: 978-1-620-917336. Gr 1-5

Informational

Adamson, Thomas K. *How Do You Measure Length and Distance?* Capstone, 2010. ISBN: 978-1-4296-4456-3. Gr PreS-2

Adler, David A. *Fractions, Decimals, and Percents*. Holiday House, 2010. ISBN: 978-0-8234-2199-2. Gr 3-5

Adler, David A. *Fun with Roman Numerals*. Holiday House, 2008. ISBN: 978-0-8234-2255-5. Gr 2 and Up

Adler, David A. *Mystery Math: A First Book of Algebra*. Holiday House, 2011. ISBN: 978-0-8234-2289-0. Gr 2-4

Adler, David A. *Perimeter, Area, and Volume*. Holiday House, 2012. ISBN: 978-0-8234-2763-5. Gr 2-5

Adler, David A. *Time Zones*. Holiday House, 2010. ISBN: 978-0-8234-2385-9. Gr 2 and Up

Arroyo, Sheri L. *How Crime Fighters Use Math*. Chelsea Clubhouse, 2009. ISBN: 978-1-60413-602-9. Gr 4-6

Ball, Johnny. *Go Figure: A Totally Cool Book about Numbers*. DK Children, 2005. ISBN: 978-075-6613-747. Gr 3-7

Ball, Johnny. *Why Pi?* DK Children, 2009. ISBN: 978-075-6651-640. Gr 2-5

Basher, Simon. *Basher Basics: A Book You Can Count On*. Kingfisher, 2010. ISBN: 978-0-75346-419-9. Gr 3-7

Brunner-Jass, Renata. *Designer Digs: Finding Area and Surface Area*. Norwood House Press, 2012. ISBN: 978-1-599-535746. Gr 5-6

Brunner-Jass, Renata. *Finding the Treasure: Coordinate Grids*. Norwood House Press, 2013. ISBN: 978-1-599-535739. Gr 3-7

Campbell, Sarah C. *Growing Patterns: Fibonacci Numbers in Nature*. Boyds Mills Press, 2010. ISBN: 978-1-59078-752-6. Gr 1-6

Campbell, Sarah C. *Mysterious Patterns: Finding Fractals in Nature*. Boyds Mills Press, 2014. ISBN: 978-1-62091-627-8. Gr 2-5

D'Agnese, Joseph. *Blockhead: The Life of Fibonacci*. Henry Holt and Co., 2010. ISBN: 978-0-8050-6305-9. Gr 4-6

D'Amico, Joan. *The Math Chef: Over 60 Math Activities and Recipes for Kids*. J. Wiley, 1997. ISBN: 978-0-4711-3813-6. Gr 4 and Up

Demuth, Patricia Brennan. *Who Is Bill Gates?* Grosset & Dunlap, 2013. ISBN: 978-0-448-463322. Gr 3-7

Dugan, Christine. *Pack It Up: Surface Area and Volume*. Teacher Created Materials, 2012. ISBN: 978-1-4333-3461-0. Gr 3-5

Einhorn, Edward. *Fractions in Disguise*. Charlesbridge, 2014. ISBN: 978-1-570-917745. Gr 2-5

Ellis, Julie. *Pythagoras and the Ratios*. Charlesbridge, 2010. ISBN: 978-1-57091-775-2. Gr 3-7

Ellis, Julie. *What's Your Angle, Pythagoras? A Math Adventure*. Charlesbridge Pub Inc., 2004. ISBN: 978-1-5709-1150-7. Gr 3 and Up

Fisher, Valorie. *How High Can a Dinosaur Count and Other Math Mysteries*. Schwartz & Wade, 2012. ISBN: 978-0-3759-3608-1. Gr 1-4

Goldsmith, Mike. *How to Be a Math Genius*. DK Children, 2012. ISBN: 978-075-6697-969. Gr 5-9

Green, Dan. *Basher Science: Algebra and Geometry*. Kingfisher, 2011. ISBN: 978-0-75346-597-4. Gr 5-9

Hanson, Anders. *Cool Optical Illusions: Creative Activities That Make Math and Science Fun for Kids*. Checkerboard Library, 2013. ISBN: 978-1-61783-822-4. Gr 3-5

Hanson, Anders. *Cool Tessellations: Creative Activities That Make Math and Science Fun for Kids*. Checkerboard Library, 2013. ISBN: 978-1-617-838262. Gr 4-6

Heos, Bridget. *At the Eleventh Hour: And Other Expressions about Money and Numbers*. Lerner Publications, 2012. ISBN: 978-0-761-381643. Gr 3-6

Heos, Bridget. *Counting Change*. Amicus, 2014. ISBN: 978-1-607-534624. Gr 1-3

Information Everywhere: The World Explained in Facts, Stats, and Graphics. Dorling Kindersley Limited, 2013. ISBN: 978-1-4654-0257-8. Gr 3 and Up

Irving, Dianne. *Volume and Hot Air Balloons*. Capstone Press, 2011. ISBN: 978-1-4296-6619-0. Gr 4 and Up

Jacoby, Jenny. *STEM Starters for Kids Math Activity Book: Packed with Activities and Math Facts*. Racehorse for Young Readers, 2017. ISBN: 978-1-631-581939. Gr 1-5

Jenkins, Steve. *Just a Second*. Houghton Mifflin, 2011. ISBN: 978-0-6187-0896-3. Gr 4-7

Kemper, Bitsy. *Budgeting, Spending, and Saving*. Lerner Publications, 2015. ISBN: 978-1-467-761055. Gr 3-5 Others

Lasky, Kathryn. *The Man Who Made Time Travel*. Farrar, Straus and Giroux, 2003. ISBN: 0-374-34788-3. Gr 4-6

Lewis, J. Patrick. *Edgar Allan Poe's Pie: Math Puzzlers in Classic Poems*. Harcourt, 2012. ISBN: 978-0-5475-1338-6. Gr 3-6

Mahaney, Ian F. *Math on the Playground: Area and Perimeter*. PowerKids Press, 2013. ISBN: 978-1-44889-657-8. Gr 3-5

Marsico, Katie. *Ball Game Math*. Lerner Publications, 2014. ISBN: 978-1-467-718851. Gr 3-4

Marsico, Katie. *Garden Math*. Lerner Classroom, 2015. ISBN: 978-1-467-786300. Gr 3-5

Marsico, Katie. *Weather Math*. Lerner Publications, 2015. ISBN: 978-1-467-786348. Gr 3-5 Other

McCallum, Ann. *Eat Your Math Homework: Recipes for Hungry Minds*. Charlesbridge Pub Inc., 2011. ISBN: 978-1-5709-1780-6. Gr 2-6

Milway, Katie Smith. *One Hen: How One Small Loan Made a Big Difference*. Kids Can Press, 2008. ISBN: 978-1-554-530281. Gr 2-5

Mooney, Carla. *Stem Jobs in Movies*. Rourke Educational Media, 2014. ISBN: 978-1-627-178235. Gr 4-8

Ottaviani, Jim. *The Imitation Game: Alan Turing Decoded*. Harry N. Abrams, 2016. ISBN: 978-1-419-7189939. YA

Perritano, John V. *Start the Game: Geometry in Sports*. Norwood House Press, 2013. ISBN: 978-1-603-575027. Gr 3-4 Others

Pickover, Clifford A. *The Math Book: From Pythagoras to the 57th Dimension, 250 Milestones in the History of Mathematics*. Sterling, 2012. ISBN: 978-1-402-788291. Gr 8 and Up

Rosen, Michael J. *Mind-Boggling Numbers*. Millbrook Press, 2016. ISBN: 978-1-467-734899. Gr 2-5

Schultz, Mark. *The Stuff of Life: A Graphic Guide to Genetics and DNA*. Hill and Wand, 2009. ISBN: 978-0-809-089475. Gr 7-12

Tang, Greg. *The Grapes of Math*. Scholastic, 2001. ISBN: 978-0-43959-840-8. Gr 2-5 Classic

Tang, Greg. *Math for All Seasons: Mind-Stretching Math Riddles*. Scholastic, 2002. ISBN: 978-0-43921-042-9. Gr 2-4

Taylor-Butler, Christine. *Understanding Charts and Graphs*. Children's Press, 2013. ISBN: 978-0-531-26240-5. Gr 2-4

Woods, Mark and Ruth Owen. *Xtreme! Extreme Sports Facts and Stats*. Gareth, 2011. ISBN: 978-1-4339-5020-9. Gr 4-6

Yang, Gene. *Secrets and Sequences*. First Second, 2017. ISBN: 978-1-626-726185. Gr 3-7

Yoder, Eric. *One Minute Mysteries: 65 Short Mysteries You Solve with Math*. Science Naturally, 2010. ISBN: 978-0-9678-0200-8. Gr 5-9

Series

A Math Adventure (Charlesbridge)

All About Counting Bugs 1-2-3 (Enslow)

Bedtime Math (Feiwel and Friends)

Brainy Day Book Series (Math Solutions Publications)

Consumer Math Series (Steck-Vaughn)

Cool Art with Math and Science (Checkerboard Library)

Core Math Skills (PowerKids Press)

Count Your Way Through . . . Afghanistan (Carolrhoda Books)

First Step Nonfiction—Graph It! (Lerner)

Hello Math Reader (Scholastic)

How Do We Use Money? (Lerner Publications)

iMath Readers (Norwood House Press)

Kumon Math Workbooks (Kumon Publishing)

Learn to Read Math Series (Creative Teaching Press)

Math Everywhere! (Lerner Books)

Math in the Real World (Chelsea)

Math Is CATegorical (Millbrook)

Math Counts (Children's Press)

Math Matters (Kane Publications)

MathStart (HarperCollins)

MathStart 2 (HarperCollins)

Money Basics (Cloverleaf Books)

Mouse Math (Kane Press)

Singapore Math (Singapore Asian Publications)

Top Score Math (Gareth)

Chapter 11

ACTIVITY BOOKS

Activities and projects are an important part of any STEAM program. These books contain carefully crafted authentic learning experiences that will provide pathways to understanding in all STEAM areas. Some activities and projects are very directed and purposeful, and others are open-ended and require planning and creativity. A complete list is provided here, divided by STEAM topic with the exception of engineering activity and project books that are located in Chapter 5: Engineering for Kids?

SCIENCE

Alderfer, Jonathan. *National Geographic Kids Bird Guide of North America: The Best Birding Book for Kids from National Geographic Bird Experts*. National Geographic Children's Books, 2013. ISBN: 978-1-426-31095-9. Gr 2-4

Burns, Loree Griffin. *Citizen Scientists: Be a Part of Scientific Discovery from Your Own Backyard*. Henry Holt, 2012. ISBN: 978-0-8050-9062-8. Gr 3-5

Cate, Annette LeBlanc. *Look Up! Bird-Watching in Your Own Backyard*. Candlewick, 2013. ISBN: 978-0-76364-561-8. Gr 2-4

Citro, Asia. *The Curious Kid's Science Book: 100+ Creative Hands-On Activities for Ages 4–8*. The Innovation Press, 2015. ISBN: 978-1-943-147007. Gr PreS-3

Citro, Asia. *A Little Bit of Dirt: 55+ Science and Art Activities to Reconnect Children with Nature*. The Innovation Press, 2016. ISBN: 978-1-943-147045. Gr PreS-3

Heinecke, Liz Lee. *Kitchen Science Lab for Kids: 52 Family Friendly Experiments from the Pantry*. Quarry Books, 2014. ISBN: 978-1592539253. Gr 2-5

Hollihan, Kerrie Logan. *Isaac Newton and Physics for Kids: His Life and Ideas with 21 Activities.* Chicago Review Press, 2009. ISBN: 978-1-556-527784. Gr 4-7

Homer, Holly. *The 101 Coolest Simple Science Experiments.* Page Street Publishing, 2016. ISBN: 978-1-624-141331. Gr 3-7

Hutchinson, Sam. *STEM Starters for Kids Science Activity Book: Packed with Activities and Science Facts.* Racehorse for Young Readers, 2017. ISBN: 978-1-631-581922. Gr 1-5

O'Quinn, Amy M. *Marie Curie for Kids: Her Life and Scientific Discoveries, with 21 Activities and Experiments.* Chicago Review Press, 2016. ISBN: 978-1-613-733202. Gr 4 and Up

Panchyk, Richard. *Galileo for Kids: His Life and Ideas 25 Activities for Kids.* Chicago Review, Press, 2005. ISBN: 978-1-556-525667. Gr 4 and Up

Pascal, Janet. *Who Was Isaac Newton?* Turtleback, 2014. ISBN: 978-0-606-361743. Gr 3-7

Pohlen, Jerome. *Albert Einstein and Relativity for Kids: His Life and Ideas with 21 Activities.* Chicago Review Press, 2012. ISBN: 978-1-613-740286. Gr 4 and Up

Robinson, Tom. *The Everything Kids' Science Experiment Book: Boil Ice, Float Water.* Everything, 2001. ISBN: 978-1-580-625579. Gr 3-7

Schulz, Karen. *CSI Expert: Forensic Science for Kids.* Prufrock Press, 2008. ISBN: 978-1-593-633127. Gr 5-8

Tomecek, Steve. *National Geographic Kids Everything Rocks and Minerals.* National Geographic Children's Books, 2011. ISBN: 978-1426307683 Gr 3-7

Wynne, Patricia J. *My First Book about Backyard Nature: Ecology for Kids.* Dover Publications, 2016. ISBN: 978-0-486-809496. Gr 3-6

Yoder, Eric. *One Minute Mysteries: 65 More Short Mysteries You Solve with Science.* Platypus Media, 2012. ISBN: 978-1-938-49200-9-6. Gr 3-7

TECHNOLOGY

Blackburn, Ken. *Kids' Paper Air Plane Book.* Workman, 1996. ISBN: 978-0-7611-0478-0. Gr 1-12

Bow, James. *Maker Projects for Kids Who Love Robotics.* Crabtree, 2016. ISBN: 978-0-778-722663. Gr 4-7

Bruzzone, Catherine. *STEM Starters for Kids Technology Activity Book: Packed with Activities and Technology Facts.* Racehorse for Young Readers, 2016. ISBN: 978-1-631-581953. Gr 1-5

Carson, Mary Kay. *Beyond the Solar System: Exploring Galaxies, Black Holes, Alien Planets, and More: A History with 21 Activities*. Chicago Review Press, 2013. ISBN: 978-1-613-74544-1. Gr 5-8

Casey, Susan. *Kids Inventing! A Handbook for Young Inventors*. Jossey-Bass, 2005. ISBN: 978-0-47-166-0-86-6. Gr 6 and Up

Ceceri, Kathy. *Robotics: Discover the Science and Technology of the Future with 20 Projects (Build It Yourself)*. Nomad Press, 2012. ISBN: 978-1-93674-975-1. Gr 3-7

DK. *3D Printing Projects*. DK, 2017. ISBN: 978-1-465-464767. Gr 4-7

Eckerson, Nate. *Stopmotion Explosion: Animate Anything and Make Movies*. Nate Eckerson, 2011. ISBN: 978-0-98333-110-0. Gr 4 and Up

Grabham, Tim, Suridh Hassan, Dave Reeve, and Clare Richards. *Movie Maker: The Ultimate Guide to Making Films*. Candlewick, 2010. ISBN: 978-0-763-649494. Gr 3-7

Harbour, Jonathan. *Video Game Programming for Kids*. Cengage Learning, 2014. ISBN: 1305501829. Gr 3-7

Isogawa, Yoshihito. *The Lego Mindstorms EV3 Idea Book: 181 Simple Machines and Clever Contraptions*. No Starch Press, 2014. ISBN: 978-1-59327-532-7. Gr 3 and Up

Kopp, Megan. *Maker Projects for Kids Who Love Electronics (Be a Maker!)*. Crabtree, 2016. ISBN: 978-0-778-725817. Gr 3-7

Marji, Majed. *Learn to Program with Scratch: A Visual Introduction to Programming with Games, Art, Science, and Math*. No Starch Press, 2014. ISBN: 978-1-59327-543-3. Gr 4 and Up

Mills, J. Elizabeth. *Creating Content: Maximizing Wikis, Widgets, Blogs, and More*. The Rosen Publishing Company, 2011. ISBN: 978-1-4488-1322-3. YA

Murphy, Maggie. *High-Tech DIY Projects with Robotics*. Powerkids Pr, 2014. ISBN: 978-1-44776-675-0. Gr 5-8

Murphy, Maggie. *High-Tech DIY Projects with 3-D Printing*. Powerkids Pr, 2014. ISBN: 978-1-47776-676-7. Gr 4 and Up

Popular Mechanics Co. *Projects for the Young Mechanic: Over 250 Classic Instructions and Plans*. Dover Publications, 2013. ISBN: 978-0-486-491172. YA

Osborne, Dave. *Woodworking Projects with and for Children*. Dave's Shop Talk from DDFM Enterprises, 2014. Kindle

Sande, Warren and Carter Sande. *Hello World! Computer Programming for Kids and Other Beginners*. Manning Publications, 2013. ISBN: 978-1-61729-092-3. YA

Shea, Therese. *Robotics Club: Teaming Up to Build Robots*. The Rosen Publishing Group, 2011. ISBN: 978-1-4488-1237-0. Gr 5 and Up

Sobey, Ed. *Electric Motor Experiments*. Enslow, 2011. ISBN: 978-0-766-033061. Gr 6-9

Sobey, Ed. *Robot Experiments*. Enslow, 2011. ISBN: 978-0-7660-3303-0. Gr 6-9

Strom, Chris. *3D Game Programming for Kids: Create Interactive Worlds with Java Script*. Pragmatic Bookshelf, 2013. ISBN: 978-1-93778-544-4. Gr 3 and Up

Van Vleet, Carmella. *Explore Electricity with 25 Great Projects*. Nomad Press, 2013. ISBN: 978-1-61930-180-1. Gr K-4

Woodcock, Jon. *Coding Games in Scratch: A Step-by-Step Visual Guide to Building Your Own Computer Games*. DK, 2016. ISBN: 978-1-465-439352. Gr 3-7

Woodcock, Jon. *Coding with Scratch Workbook*. DK, 2015. ISBN: 978-1-465-433922. Gr 1-4

Woodcock, Jon and Steve Setford. *Coding in Scratch Games Workbook*. DK, 2016. ISBN: 978-1-465-444820. Gr 1-4

ENGINEERING

Andrews, Beth L. *Hands-On Engineering: Real-World Projects for the Classroom*. Prufrock Press Inc., 2012. ISBN: 978-1-593-639228. Gr 4-7

Beaty, Andrea. *Ada Twist's Big Project Book for Stellar Scientists*. Harry N. Abrams, 2017. ISBN: 978-1-419-730245. Gr K-2

Beaty, Andrea. *Iggy Peck's Big Project Book for Amazing Architects*. Abrams, 2017. ISBN: 978-1-683-351306. Gr K-2

Beaty, Andrea. *Rosie Revere's Big Project Book for Bold Engineers*. Abrams, 2017. ISBN: 978-1-613-125304. Gr K-2

Carter, David. *The Elements of Pop-Up: A Pop-Up Book for Aspiring Paper Engineers*. Scholastic, 1999. ISBN: 0-689-82224-3. Gr 4-7

Ceceri, Kathy. *Making Simple Robots: Exploring Cutting-Edge Robotics with Everyday Stuff*. Maker Media, 2015. ISBN: 978-1-457-183638. Gr 6-12

Ceceri, Kathy and Samuel Carbaugh. *Robotics: Discover the Science and Technology of the Future with 20 Projects*. Nomad, 2012. ISBN: 978-1-936-749751 Gr 3-7

DK. *Find Out! Engineering*. DK, 2017. ISBN: 978-1-465-462343. Gr 1-4

Gurstelle, William. *The Art of the Catapult: Build Greek Ballistae, Roman Onagers, English Trebuchets, and More Ancient Artillery*. Chicago Review Press, 2004. ISBN: 978-1-55652-526-1. Gr 4 and Up

Isogawa, Yoshihito. *The Lego Technic Idea Book: Simple Machines, Book I.* Starch Press, 2010. ISBN: 978-1-59327-277-7. Gr 4 and Up

Jacoby, Jenny. *STEM Starters for Kids Engineering Activity Book: Packed with Activities and Engineering Facts.* Racehorse for Young Readers, 2017. ISBN: 978-1-631-581946. Gr 1-5

Latham, Donna. *Bridges and Tunnels: Investigate Feats of Engineering with 25 Projects.* Nomad Press, 2012. ISBN: 978-1-936-749522. Gr 3-7

Levy, Matthys. *Engineering the City How Infrastructure Works Projects and Principles for Beginners.* Paw Prints, 2008. ISBN: 978-1-43526-096-2. Gr 6 and Up

McCue, Camille. *Getting Started with Engineering: Think Like an Engineer.* John Wiley, 2016. ISBN: 978-1-119-291220. Gr 2-5

Reyes, Sandi. *Engineer through the Year: 20 Turnkey STEM Projects to Intrigue, Inspire & Challenge—Grades K-2.* SDE Crystal Springs Books, 2012. ASIN: 1935502379. Gr K-2

Reyes, Sandi. *Engineer through the Year: 20 Turnkey STEM Projects to Intrigue, Inspire & Challenge—Grades 3–5.* SDE Crystal Springs Books, 2012. ASIN: B00QM20SFC. Gr 3-5

Sobey, Ed. *The Motorboat Book: Build and Launch 20 Jet Boats, Paddle Wheelers, Electric Submarines and More.* Chicago Review Press, 2013. ISBN: 978-1-6137-4447-5. Gr 4 and Up

VanCleave, Janice. *Janice VanCleave's Engineering for Every Kid: Easy Activities That Make Learning Science Fun.* Wiley, 2007. ISBN: 978-047-1471-182-0. Gr 3 and Up

Woodroffe, David. *Making Paper Airplanes: Make Your Own Aircraft and Watch Them Fly!* Skyhorse Publishing, 2012. ISBN: 978-1-62087-168-3. Gr K-3

ART

Abel, Jessica. *Drawing Words and Writing Pictures: Making Comics: Manga, Graphic Novels and Beyond.* First Second, 2008. ISBN: 978-1-5964-3131-7. YA

Angleberger, Tom. *Art2-D2's Guide to Folding and Doodling: An Origami Yoda Activity Book.* Amulet Books, 2013. ISBN: 978-1-4197-0534-2. Gr 4 and Up

Bauer, Helen. *Beethoven for Kids: His Life and Music with 21 Activities.* Chicago Review Press, 2011. ISBN: 978-1-5697-4500-7. Gr 4 and Up

Bauer, Helen. *Verdi for Kids: His Life and Music with 21 Activities.* Chicago Review Press, 2013. ISBN: 978-1-6137-4500-7. Gr 5-8

Bidner, Jenni. *The Kids' Guide to Digital Photography: How to Shoot, Save, Play with and Print Your Digital Photos.* Lark Books, 2004. ISBN: 978-1-579-906044. Gr 4 and Up

Crossingham, John. *Learn to Speak Music: A Guide to Creating, Performing, and Promoting Your Songs.* Owlkids, 2009. ISBN: 978-1-8973-4965-6. Gr 5-8

DeCarufel, Laura. *Learn to Speak Fashion: A Guide to Creating, Showcasing & Promoting Your Style.* Owlkids Books, 2012. ISBN: 978-1-9269-7337-1. Gr 6-10

Holmes, Marc Taro. *Designing Creatures and Characters: How to Build an Artist's Portfolio for Video Games, Film, Animation and More.* North Light Books, 2016. ISBN: 978-1-440-344091. Adult

Levete, Sarah. *Maker Projects for Kids Who Love Animation.* Crabtree, 2016. ISBN: 978-0-778-722564. Gr 3-7

My Art Book: Amazing Art Projects Inspired by Masterpieces. DK, 2011. ISBN: 978-0-7566-7582-0. Gr 3-6

Neuburger, Emily. *Show Me a Story: 40 Craft Projects and Activities to Spark Children's Storytelling.* Storey Publishing, 2012. ISBN: 978-60342-988-7. Gr K-7

Roche, Art. *Art for Kids: Cartooning: The Only Cartooning Book You'll Ever Need to Be the Artist You've Always Wanted to Be.* Sterling, 2010. ISBN: 978-1-4020775154. Gr 3-12

Roche, Art. *Art for Kids: Comic Strips: Create Your Own Comic Strips from Start to Finish.* Lark Books, 2007. ISBN: 978-1-5799-0788-4. Gr 4 and Up

Salvador, Ana. *Draw with Pablo Picasso.* Frances Lincoln Children's Books, 2008. ISBN: 978-1-8450-7819-5. Gr 1 and Up

Shaskan, Trisha Speed. *Art Panels, BAM! Speech Bubbles, POW! Writing Your Own Graphic Novel.* Picture Window Books, 2011. ISBN: 978-1-4048-6393-4. Gr 2-4

Stoneham, Bill. *How to Create Fantasy Art for Video Games: Complete Guide to Creating Concepts, Characters, and Worlds.* Barron's, 2010. ISBN: 978-0-7641-4504-9. Adult

Thompson, Ruth. *The Science and Inventions Creativity Book: Games, Models to Make, High-Tech Craft Paper, Stickers, and Stencils.* Barron's Educational Series, 2013. ISBN: 978-1-438-00251-4. Gr 1-6

Watt, Fiona. *Step-by-Step Drawing Book.* Usborne Books, 2014. ISBN: 978-0-794-529536. Gr 4 and Up

Williams, Freddie. E. *The DC Comics Guide to Digitally Drawing Comics.* Watson-Guptill, 2009. ISBN: 978-0-8230-9923-8. Gr 9-12

MATH

Clemens, Meg. *The Everything Kids' Math Puzzles Book: Brain Teasers, Games, and Activities for Hours of Fun.* Everything, 2003. ISBN: 978-1-580-627733. Gr 4-6

D'Amico, Joan. *The Math Chef: Over 60 Math Activities and Recipes for Kids.* J. Wiley, 1997. ISBN: 978-0-471-138136. Gr 4-9

Hanson, Anders. *Cool Optical Illusions: Creative Activities That Make Math & Science Fun for Kids*. Checkerboard Library, 2013. ISBN: 978-1-61783-822-4. Gr 3-5

Hanson, Anders. *Cool Tessellations: Creative Activities That Make Math and Science Fun for Kids*. Checkerboard Library, 2013. ISBN: 978-1-617-838262. Gr 4-6

Jacoby, Jenny. *STEM Starters for Kids Math Activity Book: Packed with Activities and Math Facts*. Racehorse for Young Readers, 2017. ISBN: 978-1-631-581939. Gr 1-5

McCallum, Ann. *Eat Your Math Homework: Recipes for Hungry Minds*. Charlesbridge Pub Inc., 2011. ISBN: 978-1-5709-1780-6. Gr 2-6

Nelson, Robin. *Let's Make a Bar Graph*. Lerner Publications, 2013. ISBN: 978-0-7613-8972-9. Gr K-3

Nelson, Robin. *Let's Make a Tally Chart*. Lerner Publications, 2013. ISBN: 978-0-7613-8975-0. Gr K-3

Overdeck, Laura. *Bedtime Math: A Fun Excuse to Stay Up Late*. Feiwel and Friends, 2013. ISBN: 978-1-250-03595-1. Gr PreS-2

Overdeck, Laura. *Bedtime Math 2: This Time It's Personal*. Feiwel and Friends, 2014. ISBN: 978-1-250-040961. Gr PreS-4. Others

Tang, Greg. *The Grapes of Math*. Scholastic, 2001. ISBN: 978-0-43959-840-8. Gr 2-5 Classic

Tang, Greg. *Math for All Seasons: Mind-Stretching Math Riddles*. Scholastic, 2002. ISBN: 978-0-43921-042-9. Gr 2-4

Yoder, Eric. *One Minute Mysteries: 65 Short Mysteries You Solve with Math*. Science Naturally, 2010. ISBN: 978-0-9678-0200-8. Gr 5-9

GENERAL

Burker, Josh. *The Invent to Learn Guide to Fun Classroom Technology Projects*. Constructing Modern Knowledge Press, 2015.

Butz, Steve. *Year-Round Project-Based Activities for STEM Grades 2-3*. Teacher Created Resources, 2013.

Cameron, Schyrlet and Carolyn Craig. *Project-Based Learning Tasks for Common Core State Standards: Solving Real-Life Problems Gr. 6-8*. Mark Twain Media, 2014.

Cameron, Schyrlet and Carolyn Craig. *STEM Labs for Middle Grades: 50+ Integrated Labs*. Mark Twain Media, 2016.

Carey, Anne. *STEAM Kids: 50+ Science/Technology/Engineering/Art/Math Hands-On Projects for Kids*. CreateSpace Independent Publishing, 2016.

Ceceri, Kathy. *Making Simple Robots: Exploring Cutting-Edge Robotics with Everyday Stuff.* Maker Media, Inc., 2015.

Challoner, Jack. *Maker Lab: 28 Super Cool Projects: Build * Invent * Create * Discover.* DK Children, 2016.

Cline, Lydia Sloan. *3D Printer Projects for Makerspaces.* McGraw-Hill Education, 2017.

Doorley, Rachelle. *Tinkerlab: A Hands-On Guide for Little Inventors—55 Playful Experiments That Encourage Tinkering, Curiosity and Creative Thinking.* Roost Books, 2014.

Drumm, Brooke, James Floyd Kelly, John Edgar Park, John Baichtal, Brian Roe, Nick Ernst, Steven Bolin, and Caleb Cotter. *3D Printing Projects: Toys, Bots, Tools, and Vehicles to Print Yourself.* Maker Media, 2015.

Frinkle, Andrew. *50 More STEM Labs: Science Experiments for Kids.* CreateSpace Independent Publishing, 2014.

Frinkle, Andrew. *50 STEAM Labs: Thematic Projects: Science, Technology, Engineering, Art, and Math.* CreateSpace Independent Publishing, 2017.

Frinkle, Andrew. *50 STEM Labs: Science Experiments for Kids.* CreateSpace Independent Publishing, 2014.

Gabrielson, Curt. *Make: Tinkering: Kids Learn by Making Stuff.* Maker Media, 2015.

Graves, Colleen and Aaron Graves. *The Big Book of Maker Space Projects.* McGraw-Hill Education, 2016.

Hand, Jamie and Amanda Boyarshinov. *STEAM: Preschool Activities for STEM Enrichment.* Kindle eBook, 2014.

Kurowski, Kathryn. *Year-Round Project-Based Activities for STEM Grades Pre-K-K.* Teacher Created Resources, 2013.

Lester, Stephanie. *Year-Round Project-Based Activities for STEM Grades 1-2.* Teacher Created Resources, 2013.

Martin, Danielle and Alisha Panjwani. *Start Making! A Guide to Engaging Young People in Maker Activities.* Edited by Natalie Rusk. Maker Media, 2016.

McCue, Camille. *Getting Started with Engineering: Think Like an Engineer.* (Dummies Junior). For Dummies, 2016.

Mukherjee, Sumita. *STEAM Ahead! DIY for Kids: Activity Pack with Science/Technology/Engineering/Art/Math Making in Building Activities for 4 to 10-Year-Old Kids.* CreateSpace Independent Publishing, 2016.

Powers, Michelle, Teri Barenborg, Tari Sexton, and Lauren Monroe. *STEAM Design Challenges*. Creative Teaching Press, 2017.

Reyes, Sandi. *Engineer through the Year: 20 Turnkey STEM Projects to Intrigue, Inspire and Challenge—Grades K-2*. SDE Crystal Springs Books, 2012.

Reyes, Sandi. *Engineer through the Year: 20 Turnkey STEM Projects to Intrigue, Inspire and Challenge—Grades 3–5*. SDE Crystal Springs Books, 2012.

Traig, Jennifer, Ed. *STEM to Story: Enthralling and Effective Lesson Plans for Grades Five through Eight*. Jossey-Bass, 2015.

VanCleave, Janice. *Janice VanCleave's Engineering for Every Kid: Easy Activities That Make Learning Science Fun*. Wiley, 2007.

Wolpert-Gawron, Heather. *DIY Project Based Learning for ELA and History*. Routledge, 2015.

Wolpert-Gawron, Heather. *DIY Project Based Learning for Math and Science*. Routledge, 2016.

Appendix A

WEBSITES

MAKERSPACES

Design Squad Global—http://pbskids.org/designsquad/parentseducators/

Do It Yourself (DIY)—https://diy.org/guides/classrooms—awesome skills for awesome kids

Drawdio—https://drawdio.com/—a pencil that lets you draw music

Global Cardboard Challenge—http://imagination.org/our-projects/cardboard -challenge/

Invent to Learn—https://inventtolearn.com/

Kids Gardening—https://kidsgardening.org/

LittleBits—http://littlebits.cc/—droid kits

Make It @ Your Library—http://makeitatyourlibrary.org/

Maker Camp—https://makercamp.com/—archived projects

Maker Education Initiative—http:/makered.org

Maker-State—http://www.maker-state.com

Makey Makey—https://www.makeymakey.com/—great kits for $50

Making Apps—http://appinventor.mit.edu/explore/

Picoboards—http://www.picocricket.com/

Science Toy Maker—http://www.sciencetoymaker.org/

Scratch—https://scratch.mit.edu/—create stories, games, and animations

Silvia's Super Awesome Maker Show—http://sylviashow.com/

SparkFun—https://learn.sparkfun.com/—electronics

Squishy Circuits—http://squishycircuits.com/

STEAM-Makers—http://www.steam-makers.com/

3D PRINTING

3D Ponics—https://www.3dponics.com/

Instructables—http://www.instructables.com/

My Mini Factory—https://www.myminifactory.com/

Pinshape—https://pinshape.com/

SketchUp—http://www.sketchup.com/

Thingverse—https://www.thingiverse.com/—3D Designs

TinkerCAD—https://www.tinkercad.com/—create 3D designs

Yobi3D—https://www.yobi3d.com/—search engine with links to 3D designs

YouMagine—https://www.youmagine.com/

MAKER EDUCATION

Edutopia Makers Area—https://www.edutopia.org/topic/maker-education resources

Exploratorium—https://www.exploratorium.edu/

Genius Hour—http://www.geniushour.com/

How Stuff Works—https://www.howstuffworks.com/

Institute of Play—https://www.instituteofplay.org/—transforming education through play

The Kids Should See This—http://thekidshouldseethis.com/—smart videos for curious minds of all ages

Odyssey of the Mind—https://www.odysseyofthemind.com/p/

PBL WEBSITES

BEANZ: The Magazine for Kids, Code, and Computer Science—https://www.kidscodecs.com/

Global School Net—http://www.globalschoolnet.org/gsnpr/

IEARN—collaborative project work worldwide—https://iearn.org/country/iearn-usa

Mathalicious—http://www.mathalicious.com/

More Project-Based Learning—http://archive.pbl-online.org/

Next Lesson—https://www.nextlesson.org/

Project-Based Learning—https://21centuryedtech.wordpress.com

We Are Teachers—https://www.weareteachers.com

ENGINEERING WEBSITES

100 Engineering Projects for Kids—https://thehomeschoolscientist.com/100-engineering-projects-kids/

Children's Engineering—http://childrensengineering.org/

Children's Engineering Design Briefs—http://childrensengineering.org/design-briefs/

Cooper Hewitt Design Lesson Plans—http://dx.cooperhewitt.org/lesson-plans/

Discover Engineering—http://www.discovere.org/discover-engineering

EdHeads—http://edheads.org/

Engineer Girl—https://www.engineergirl.org/

Engineering Adventures—https://eie.org/engineering-adventures

Engineering and Technology History Wiki—http://ethw.org/Main_Page

Engineering Everywhere—https://eie.org/engineering-everywhere

Engineering for Kids—http://www.sciencekids.co.nz/engineering.html

Engineering Is Elementary—https://eie.org/

Engineering: Go for It—http://www.egfi-k12.org/

International Technology and Engineering Educators Association—https://www.iteea.org/

PBS Kids Design Squad Global—http://pbskids.org/designsquad/

SPARK!LAB—http://invention.si.edu/current-sparklab-activities

Stem Village—http://www.stemvillage.com

Teach Engineering—https://www.teachengineering.org/

The Ultimate STEM Guide for Kids: 239 Cool Sites—http://www.mastersindatascience.org/blog/the-ultimate-stem-guide-for-kids-239-cool-sites-about-science-technology-engineering-and-math/

Try Engineering—http://tryengineering.org/home

Try Engineering Games—http://tryengineering.org/play-games

What's an Engineer? Crash Course Kids—https://www.youtube.com/watch?v=owHF9iLyxic

Appendix B

BOOKS THAT PROMOTE MIND GROWTH

Barnes, Derrick. *Crown: An Ode to the Fresh Cut*. Agate/Bolden, 2017. ISBN: 978-1-572-842243. Gr K-3

Beaty, Andrea. *Rosie Revere, Engineer*. Harry N. Abrams, 2013. ISBN: 978-1-41970-845-9. Gr K and Up

Bridges, Shirin Yim. *Ruby's Wishes*. Chronicle Books, 2015. ISBN: 978-1-452-145693. Gr K-3

Bryant, Jen. *A Splash of Red: The Life and Art of Horace Pippin*. Knopf Books for Young Readers, 2013. ISBN: 978-0-3758-6712-5. Gr K-3

Clinton, Chelsea. *She Persisted*. Philomel Books, 2017. ISBN: 978-1-524-741723. Gr PreS-3

Cook, Julia. *Bubble Gum*. National Center for Youth Issues, 2017. ISBN: 978-1-937870430. Gr PreS-3

Cook, Julia. *Thanks for the Feedback*. Boys Town Press, 2013. ISBN: 978-1-934-490495. Gr K-3

Coulman, Valerie. *When Pigs Fly*. Lobster Press, 2001. ISBN: 978-1-894-222365. Gr PreS and Up

Deak, Joann. *Your Fantastic Elastic Brain: Stretch It, Shape It*. Little Pickle Press, 2010. ISBN: 978-0-982-993804. Gr PreS-3

Demi. *The Empty Pot*. Square Fish, 1996. ISBN: 978-0-805-049008. Gr 1-5

Dr. Seuss. *Horton Hears a Who*. Random House Books for Young Readers, 1954. ISBN: 978-0-394-800783. Gr K-4

Engle, Margarita. *Drum Dream Girl*. HMH Books for Young Readers, 2015. ISBN: 978-0-544-102293. Gr PreS-3

Gray, Karlin. *Nadia, the Girl Who Couldn't Sit Still*. HMH Books for Young Readers, 2016. ISBN: 978-0-544-319608. Gr 1-4

Henkes, Kevin. *Owen*. Greenwillow Books, 1993. ISBN: 978-0-688-114497. Gr PreS-3

Howe, James. *Brontorina*. Candlewick, 2013. ISBN: 978-0-763-653231. Gr PreS-2

Jeffers, Oliver. *How to Catch a Star*. Philomel Books, 2004. ISBN: 978-0-399-242861. Gr PreS-2

Jordan, Deloris. *Salt in His Shoes*. Simon & Schuster, 2003. ISBN: 978-0-689-834196. Gr PreS-3

Judge, Lita. *Flight School*. Atheneum Books for Young Readers, 2014. ISBN: 978-1-442-481770. Gr PreS-3

Keats, Ezra Jack. *Whistle for Willie*. Puffin Books, 1991. ISBN: 978-0-140-502022. Gr PreS-2

Lewis, Kevin. *My Truck Is Stuck*. Disney-Hyperion, 2002. ISBN: 978-0-786-805341. Gr PreS-K

Luyken, Corinna. *The Book of Mistakes*. Dial Books, 2017. ISBN: 978-0-735-227927. Gr PreS-3

Pett, Mark. *The Girl Who Never Made Mistakes*. Sourcebooks Jabberwocky, 2011. Gr PreS-3

Piper, Watty. *The Little Engine That Could*. Grosset & Dunlap, 2001. ISBN: 978-0-448-405209. Gr PreS-2

Portis, Antoinette. *Not a Box*. HarperCollins, 2006. ISBN: 978-0-061-123221. Gr PreS and Up

Raschka, Chris. *Everyone Can Learn to Ride a Bicycle*. Schwartz & Wade, 2013. ISBN: 978-0-375-870071. Gr PreS-3

Reiley, Carol E. *Making a Splash*. Go Brain, 2015. ISBN: 978-0-986-417306. Adult

Reynolds, Peter H. *The Dot*. Candlewick, 2003. ISBN: 978-0-7636-1961-9. Gr PreS-4

Reynolds, Peter H. *Ish*. Walker Books Ltd., 2005. ISBN: 978-1-844-282968. Gr PreS-2

Rosenthal, Amy Krouse. *The OK Book.* HarperCollins, 2007. ISBN: 978-0-061-152559. Gr PreS-3

Rubin, Susan Goldman. *Maya Lin.* Chronicle, 2017. ISBN: 978-1-452-108377. Gr 4-8

Saltzberg, Barney. *Beautiful Oops!* Workman Publishing Company, 2010. ISBN: 978-0-761-157281. Gr PreS-2

Santat, Dan. *After the Fall: How Humpty Dumpty Got Up Again.* Roaring Brook, 2017. ISBN: 978-1-626-726826. Gr K and Up

Smith, Cynthia Leitich. *Jingle Dancer.* HarperCollins, 2000. ISBN: 978-0-688-162412. Gr PreS-5

Spires, Ashley. *The Most Magnificent Thing.* The Kids Can Press, 2014. ISBN: 978-1-554-537044. Gr PreS-2

Spires, Ashley. *The Thing Lou Couldn't Do.* Kids Can Press, 2017. ISBN: 978-1-771-387279. Gr PreS-2

Steig, William. *Brave Irene.* Square Fish, 2011. ISBN: 978-0-312-564223. Gr PesS-3

Stevens, Janet. *The Tortoise and the Hare.* Holiday House, 1984. ISBN: 978-0-823-405640. Gr K-2

Thompson, Laurie Ann. *Emmanuel's Dream: The True Story of Emmanuel Ofosu Yeboah.* Schwartz & Wade, 2015. ISBN: 978-0-449-817445. Gr PreS-3

Uegaki, Chieri. *Hana Hashimoto, Sixth Violin.* Kids Can Press, 2014. ISBN: 978-1-894-786331. Gr PreS-3

Yamada, Kobi. *What Do You Do with an Idea?* Compendium Inc., 2014. ISBN: 978-1-938-298073. Gr K-3

Yamada, Kobi. *What Do You Do with a Problem?* Compendium Inc., 2016. ISBN: 978-1-943- 200009. Gr PreS-3

Appendix C

PROFESSIONAL RESOURCES

Anderson, Chris. *Makers: The New Industrial Revolution.* Crown Business, 2014. ISBN: 978-0-307-720962.

Bender, William. *Project-Based Learning: Differentiating Instruction for the 21st Century.* Corwin, 2012. ISBN: 978-1-412-997904.

Boss, Suzie. *Implementing Project-Based Learning.* Solution Tree, 2015. ISBN: 978-1-942-496113.

Bybee, Rodger W. *The Case for STEM Education: Challenges and Opportunities.* National Science Teachers Association, 2013. ISBN: 978-1-936-959259.

Cano, Lesley M. *3D Printing: A Powerful New Curriculum Tool for Your School Library.* Libraries Unlimited, 2015. ISBN: 978-1-610-699778.

Clapp, Edward P., Jessica Ross, Jennifer O. Ryan, and Shari Tishman. *Maker-Centered Learning: Empowering Young People to Shape Their Worlds.* Jossey-Bass, 2016. ISBN: 978-1-119-259701.

"Classroom Playbook" Maker Faire. Maker Media, May 2016. Online: http://yydxg3i41b1482qi9hidybgs-wpengine.netdna-ssl.com/wp-content/uploads/2015/03/MakerFaireClassPlaybook-Apr2016.pdf

Cooper, Ross and Erin Murphy. *Hacking Project Based Learning: 10 Easy Steps to PBL and Inquiry in the Classroom.* Times 10 Publications, 2016. ISBN: 978-0-998-570518.

Dougherty, Dale. *Free to Make: How the Maker Movement Is Changing Our Schools, Our Jobs, and Our Minds.* North Atlantic Books, 2016. ISBN: 978-1-623-170745.

Elliott, Lori. *Project Based Learning for Real Kids and Real Teachers.* SDE, 2016. ISBN: 978-1-631-330864.

Farber, Katy. *Real and Relevant: A Guide for Service and Project-Based Learning.* Rowman & Littlefield, 2017. ISBN: 978-1-475-835458.

Fehrenbacher, Tom and Randy Scherer. *Hands and Minds: A Guide to Project-Based Learning for Teachers by Teachers.* CreateSpace Independent Publishing Platform, 2017. ISBN: 978-1-548-297616.

Fleming, Laura. *The Kickstart Guide to Making Great Makerspaces.* Corwin, 2017. ISBN: 978-1-506-392523.

Fleming, Laura. *Worlds of Making: Best Practices for Establishing a Makerspace for Your School.* Corwin, 2015. ISBN: 978-1-183-382821.

France, Anna Kaziunas and Brian Jepson. *Make: 3D Printing: The Essential Guide to 3D Printers.* Maker Media, 2013. ISBN: 978-1-457-182938.

Johnson, Carla C., Erin E. Peters-Burton, and J. Moore. *STEM Road Map: A Framework for Integrated STEM Education.* Routledge, 2015. ISBN: 978-1-138-804234.

Kloski, Liza and Nick Kloski. *Make: Getting Started with 3d Printing: A Hands-On Guide to the Hardware, Software, and Services behind the New Manufacturing Revolution.* O'Reilly & Associates Inc., 2016. ISBN: 978-1-680-450200.

Kraus, Jane and Suzie Boss. *Thinking through Project-Based Learning: Guiding Deeper Inquiry.* Corwin, 2013. ISBN: 978-1-452-202563.

Lang, David. *Zero to Maker: A Beginner's Guide to the Skills, Tools, and Ideas of the Maker Movement.* Maker Media, 2017. ISBN: 978-1-680-453416.

Larmer, John, John R. Mergendoller, and Suzie Boss. *Setting the Standard for Project Based Learning: A Proven Approach to Rigorous Classroom Instruction.* ASCD, 2015. ISBN: 978-1-416-620334.

Laur, Dayna and Jill Ackers. *Developing Natural Curiosity through Project-Based Learning: Five Strategies for the PreK-3 Classroom.* Routledge, 2017. ISBN: 978-1-138-694217.

"Maker Club Playbook" Young Makers. Maker Ed, January 2012.Online: http://makered.org/wp-content/uploads/2016/10/Maker-Club-Playbook_Young-Makers-Jan-2012-6_small.pdf

"Makerspace Playbook School Edition" Makerspace. Maker Media, Spring 2013. Online: https://makered.org/wp-content/uploads/2014/09/Makerspace-Playbook-Feb-2013.pdf

Martinez, Sylvia Libow and Gary S. Stager. *Invent to Learn: Making, Tinkering, and Engineering in the Classroom.* Constructing Modern Knowledge Press, 2013. ISBN: 978-0-997-554328.

Maslyk, Jacie. *STEAM Makers: Fostering Creativity and Innovation in the Elementary Classroom.* Corwin, 2016. ISBN: 978-1-506-311241.

McDowell, Michael P. *Rigorous PBL by Design: Three Shifts for Developing Confident and Competent Learners.* Corwin, 2017. ISBN: 978-1-506-359021.

Otfinoski, Steven. *3D Printing: Science, Technology, Engineering.* Children's Press, 2017. ISBN: 978-0-531-219881.

Peppler, Kylie A., Erica Halverson, and Yasmin B. Kafai. *Makeology: Makerspaces as Learning Environments.* Routledge, 2016. ISBN: 978-1-138-847774.

Riley, Susan M. *No Permission Required: Bringing STEAM to Life in K-12 Schools.* Visionyst Press, 2014. ISBN: 978-0-692-026571.

Riley, Susan M. *STEAM Point: A Guide to Integrating Science, Technology, Engineering, the Arts and Math through Common Core.* Visionyst Press, 2012. ISBN: 978-1-481-165730.

Sousa, David A. and Thomas Pilecki. *From STEM to STEAM: Using Brain-Compatible Strategies to Integrate the Arts.* Corwin, 2013. ISBN: 978-1-452-258331.

Spencer, John. *Launch: Using Design Thinking to Boost Creativity and Bring Out the Maker in Every Student.* Dave Burgess Consulting, Inc., 2016. ISBN: 978-0-996-989657.

Thornburg, David D., Norma Thornburg, and Sara Armstrong. *The Invent to Learn Guide to 3D Printing in the Classroom: Recipes for Success.* Constructing Modern Knowledge Press, 2014. ISBN: 978-0-989-151146.

Warren, Acacia M. *Project-Based Learning across the Disciplines: Plan, Manage, and Assess through +1 Pedagogy.* Corwin, 2016. ISBN: 978-1-506-333793.

WORKS CITED

Abrams, Michael. "Bot for Tots." *American Society of Mechanical Engineers*. February 2014. Accessed July 10, 2017. https://www.asme.org/career-education/articles/k-12-grade/bots -for-tots

Adams, Caralee. "How to Use a Desktop 3D Printer in School." *WeAreTeachers*. July 17, 2017. Accessed August 4, 2017. https://www.weareteachers.com/the-non-tech-teacher-s -guide-to-using-a-3d-printer/

Anglin, Nick. "A Student Maker and the Birth of a Startup." *Edutopia*. August 27, 2015. Accessed July 27, 2017. https://www.edutopia.org/blog/student-maker-birth-of-startup -nick-anglin

Bajarin, Tim. "Maker Faire: Why the Maker Movement Is Important to America's Future." *Time*. May 19, 2014. Accessed May 24, 2017. http://time.com/104210/maker-faire -maker-movement/

Bentley, Kipp. "Makerspaces: New Prospects for Hands-On Learning in Schools." *Center for Digital Education and Converge Magazine*. February 12, 2017. Accessed July 27, 2017. http://www.centerdigitaled.com/blog/makerspaces.html

Berg, Brenda. "The Critical Importance of STEAM Education." *PC World*. June 27, 2017. Accessed May 24, 2017. https://www.pcworld.idg.com.au/article/621170/critical- importance-steam-education/

Berkowicz, Jill and Ann Myers. "How a STEM Shift Makes Way for Equity." *Corwin Connect*. November 29, 2016. Accessed May 24, 2017. http://corwin-connect.com/2016/ 03/stem-shift-makes-way-equity/

Borovoy, Amy. "5-Minute Film Festival: Learning with Rube Goldberg Machines." *Edutopia*. June 24, 2016. Accessed July 10, 2017. https://www.edutopia.org/film-fest-rube-goldberg-learning-ideas

Busch, Laura. "How Should We Measure the Impact of Makerspaces?" *EdSurge News*. May 31, 2017. Accessed July 27, 2017. https://www.edsurge.com/news/2017-01-09-how-should-we-measure-the-impact-of-makerspaces

Carbon, Kim. "Going from STEM to STEAM, Transforming Education with 3D Printing." *364-Type-A Machines*. April 21, 2017. Accessed August 4, 2017. https://www.typeamachines.com/blog/stem-to-steam-transforming-education-with-3d-printing

Carmody, Erin. "Running an Engineering Design Challenge—5 Tips to Get Anyone Started." *STEM Village*. October 12, 2016. Accessed July 10, 2017. http://www.stemvillage.com/running-an-engineering-design-challenge-5-tips-to-to-get-anyone-started/

Cummins, Sunday. "Reading about Real Scientists." *Educational Leadership* 72:4 (2014–2015): 68–72.

Cunningham, Christine M. and Melissa Higgins. "Engineering for Everyone." *Educational Leadership* 72:4 (2014–2015): 42–47.

Dillon, P. Mathew. "Makerspace Technology: Is It Right for Your School?" *Edutopia*. January 31, 2017. Accessed July 27, 2017. https://www.edutopia.org/discussion/makerspace-technology-it-right-your-school

Duke, Nell K. and Anne-Lise Halvorsen. "New Study Shows the Impact of PBL on Student Achievement." *Edutopia*. June 20, 2017. Accessed June 5, 2017. https://www.edutopia.org/article/new-study-shows-impact-pbl-student-achievement-nell-duke-anne-lise-halvorsen

"Elementary Engineering: From Simple Machines to Life Skills." *Edutopia*. January 26, 2016. Accessed July 10, 2017. https://www.edutopia.org/practice/elementary-engineering-simple-machines-life-skills

Finley, Todd. "Jaw-Dropping Classroom 3D Printer Creations." *Edutopia*. June 30, 2015. Accessed August 4, 2017. https://www.edutopia.org/blog/jaw-dropping-classroom-3d-printer-todd-finley

Fleming, Laura. "Out of the Box Approach to Planning Makerspaces." *Worlds of Learning*. April 4, 2017. Accessed July 27, 2017. https://worlds-of-learning.com/2017/04/04/box-approach-makerspaces/

Fleming, Laura and Billy Krakower. "Makerspaces and Equal Access to Learning." *Edutopia*. July 19, 2016. Accessed July 27, 2017. https://www.edutopia.org/blog/makerspaces-equal-access-to-learning-laura-fleming-billy-krakower

Fleming, Laura and Ross Cooper. "Makerspace Stories and Social Media: Leveraging the Learning." *Edutopia*. September 1, 2016. Accessed July 27, 2017. https://www.edutopia.org /blog/makerspace-social-media-leveraging-learning-ross-cooper-laura-fleming

Gidcumb, Brianne. "PBL and STEAM: Do They Intersect?" *Education Closet*. October 29, 2016. Accessed June 5, 2017. https://educationcloset.com/2014/05/23/pbl-and-steam -where-do-they-intersect/

Gorman, Dorothy Powers. "Engineering in Elementary Schools: Engaging the Next Generation of Problem Solvers Today." *South East Education Network* 16:3 (2014): 36–37.

Gorman, Michael. "Part 1: STEM, STEAM, Makers: Connecting Project Based Learning (PBL)." *21st Century Educational Technology and Learning*. December 22, 2016. Accessed June 5, 2017. https://21centuryedtech.wordpress.com/2016/07/05/part-1-stem-steam -makers-connecting-project-based-learning-pbl/

Gorman, Michael. "What Is STEM Noun or Verb? STEM in All Areas . . .. Ten Ideas to Transform STEM from Nouns to Verbs . . . and Facts to Thinking." *21st Century Tech*. September 26, 2016. Accessed May 24, 2017. https://21centuryedtech.wordpress.com /2016/09/26/stem-for-in-all-areas-ten-ideas-to-transform-stem-from-nouns-to-verbs -and-facts-to-thinking/

Hale, Brent. "175 Amazing Ways 3D Printing Is Changing the World." *3D Forged*. March 27, 2017. Accessed August 4, 2017. https://3dforged.com/3d-printing/

Harper, Charlie. "The STEAM-Powered Classroom." *Educational Leadership* 75 (2017): 70–74.

Herold, Benjamin. "Researchers Probe Equity, Design Principles in Maker Ed." *Education Week* 35 (2016): 8–9.

Jarrett, Kevin. "Digital Shop Class: Fun and Profitable." *Edutopia*. January 5, 2017. Accessed July 27, 2017. https://www.edutopia.org/article/digital-shop-class-fun-profitable -kevin-jarrett

Jolly, Anne. "The Search for Real-World STEM Problems." *Education Week Teacher*. July 21, 2017. Accessed June 5, 2017. http://www.edweek.org/tm/articles/2017/07/17/the -search-for-real-world-stem-problems.html

Kane, Karen. "9 Ways Teachers Can Use a 3D Printer to Teach Math and Science." *WeAreTeachers*. February 24, 2017. Accessed June 5, 2017. https://www.weareteachers .com/3d-printing-math-science/

Larmer, John. "It's a Project-Based World." *Educational Leadership* 73:6 (2016): 66–70.

Larmer, John and John Mergendoller. "Seven Essentials for Project-Based Learning." *Educational Leadership*. Accessed June 5, 2017. http://www.ascd.org/publications/

educational_leadership/sept10/vol68/num01/Seven_Essentials_for_Project-Based
_Learning.aspx

Markham, Thom. "STEM, STEAM, and PBL." *ASCD Edge*. 2012. Accessed June 5,
2017. http://edge.ascd.org/blogpost/stem-steam-and-pbl

Martinez, Sylvia. "Making for All: How to Build an Inclusive Makerspace." *EdSurge News*.
May 10, 2015. Accessed July 27, 2017. https://www.edsurge.com/news/2015-05-10
-making-for-all-how-to-build-an-inclusive-makerspace

Mersand, Shannon. "What's New in 3D Printing." *Tech Learning*. September 2017.
Accessed October 10, 2017. http://www.techlearning.com/resources/0003/whats-new-in
-3d-printing/70729

Meyer, Leila. "7 Tips for Planning a Makerspace." *The Journal*. February 23, 2017.
Accessed July 27, 2017. https://thejournal.com/articles/2017/02/23/7-tips-for-planning
-a-makerspace.aspx

Miller, Andrew. "In Search of the Driving Question." *Edutopia*. August 30, 2017. Accessed
October 10, 2017. https://www.edutopia.org/article/search-driving-question

Miller, Andrew. "PBL and STEAM Education: A Natural Fit." *Edutopia*. May 20, 2014.
Accessed June 5, 2017. https://www.edutopia.org/blog/pbl-and-steam-natural-fit-andrew
-miller

Murphy, Kate. "UC Students Build 3D-Printed Prosthetic Hands for Kids." *USA Today*.
February 16, 2017. Accessed August 4, 2017. https://www.usatoday.com/story/tech/
nation-now/2017/02/17/students-build-3-d-printed-prosthetic-hands-kids/
98071192/

Newell, Jennifer. "Girls, Computers, and STEAM." *BEANZ*. July 31, 2017. Accessed
May 24, 2017. https://www.kidscodecs.com/girls-computers-steam/

O'Brien, Chris. "Makerspaces Lead to School and Community Successes." *Edutopia*.
March 21, 2016. Accessed July 27, 2017. https://www.edutopia.org/blog/makerspaces
-school-and-community-successes-chris-obrien

Slavin, Tim. "What Is 3D Printing? Kids, Code, and Computer Science." *BEANZ*.
January 31, 2015. Accessed August 4, 2017. https://www.kidscodecs.com/what-is-3d
-printing/

Soule, Helen. "Why STEAM Is Great Policy for the Future of Education." *P21.org*.
March 31, 2016. Accessed May 24, 2017. http://www.p21.org/news-events/p21blog/
1900-why-steam-is-great-policy-for-the-future-of-education

Spencer, John. "Why Every Classroom Should Be a Makerspace." *John Spencer*. July 19,
2017. Accessed July 27, 2017. http://www.spencerauthor.com/classroom-makerspace/

Stager, Gary S. "Unconventional Wisdom about the Maker Movement." *Invent to Learn.* Winter, 2017. Accessed July 27, 2017. https://inventtolearn.com/unconventional-wisdom-about-the-maker-movement/

Stager, Gary S. "What's the Maker Movement and Why Should I Care?" *Scholastic Administrator.* Winter 2014. Accessed May 24, 2017. http://www.scholastic.com/browse/article.jsp?id=3758336.

"STEAM + Project-Based Learning: Real Solutions from Driving Questions." *Edutopia.* January 26, 2016. Accessed June 5, 2017. https://www.edutopia.org/practice/steam-project-based-learning-real-solutions-driving-questions

Stewart, Louise. "Maker Movement Reinvents Education." *Newsweek.* September 8, 2014. Accessed May 24, 2017. http://www.newsweek.com/2014/09/19/maker-movement-reinvents-education-268739.html

Tahnk, Jeana Lee. "6 Awesome Ways to Bring Your Kids' Ideas to Life with 3D Printing." *Mashable.* September 19, 2015. Accessed August 4, 2017. http://mashable.com/2015/09/19/kids-toys-3d-printing/

Thornburg, David D. "The 3D Printing Revolution in Education." *ESchool News.* January 29, 2016. Accessed August 4, 2017. https://www.eschoolnews.com/whitepapers/the-3d-printing-revolution-in-education/

"3D Printing Educator Spotlight On: Jayda Pugliese, 5th Grade Teacher with a STEAM-Focused Classroom, Philadelphia." *3DPrint.com.* July 7, 2017. Accessed August 4, 2017. https://www.3dprint.com/180018/educator-spotlight-jayda-pugliese/

"3D Printing Educator Spotlight On: Megan Finesilver, 2nd Grade Teacher, Speaker and Curriculum Developer, Colorado." *3DPrint.com.* July 17, 2017. Accessed August 4, 2017. https://www.3dprint.com/181092/spotlight-megan-finesilver/

Ullman, Ellen. "Making the Grade: How Schools Are Creating and Using Makerspaces." *Tech Learning.* March 24, 2016. Accessed July 27, 2017. http://www.techlearning.com/resources/0003/making-the-grade-how-schools-are-creating-and-using-makerspaces/69967

Ullman, Ellen. "How Schools Make 'Making' Meaningful." *Tech Learning.* September 2017. Accessed October 10, 2017. http://www.techlearning.com/resources/0003/how-schools-make-making-meaningful/70727

Vasquez, Jo Anne. "STEM—Beyond the Acronym." *Educational Leadership* 72:4 (2014–2015): 10–15.

Wasserman, Todd. "Intersection of Creativity, Technology and Learning: A Conversation." *Worlds of Learning.* October 26, 2017. Accessed November 1, 2017. https://worlds-of-learning.com/2017/10/26/intersection-creativity-technology-learning-conversation/

Waters, Patrick. "Project-Based Learning through a Maker's Lens." *Edutopia*. July 9, 2014. Accessed June 5, 2017. https://www.edutopia.org/blog/pbl-through-a-makers-lens -patrick-waters

Wilkinson, Karen, Bevan Bronwyn, and Mike Petrich. "Tinkering Is Serious Play." *Educational Leadership* 72:4 (2014–2015): 28–33.

Wolfe, Aiden. "How to Turn Any Classroom into a Makerspace." *Edudemic*. April 9, 2015. Accessed July 27, 2017. http://www.edudemic.com/turn-classroom-makerspace/

Wright, Tanya and Amelia Wenk Gotwals. "Supporting Disciplinary Talk from the Start of School: Teaching Students to Think and Talk Like Scientists." *The Reading Teacher* 71:2 (September/October 2017): 189–197.

AUTHOR INDEX

TITLE INDEX

About the Authors

LIZ KNOWLES, EdD, is owner/director of Cognitive Advantage (www .cognitive-fitness.com), a cognitive skill development program for all ages. She has authored a differentiated reading book and has coauthored 10 books with Martha Smith. She writes curriculum for Palm Beach County Schools and serves as an educational consultant for private schools in the area. She has also taught grades K-6 and served as director of Professional Development and Curriculum for a large private school, adjunct professor in graduate reading programs at two universities, and Head of Studies at an international baccalaureate school.

MARTHA SMITH, MSE, received her undergraduate degree in library science from Eastern Michigan University and her master's degree in education in library science from the University of South Florida. She has been a media specialist for over 30 years in the pre-K through eighth grades, during which time she planned, organized, and oversaw collection development for and managed the project of building two school libraries.

Martha and Liz have coauthored 10 books on various topics of interest to teachers and librarians and were the recipients of the International Reading Association Middle School Special Interest Group Award, 2003.